He Gave Me Back My Life

Kenneth Pride Sr.

Fulton Books
Meadville, PA

Published by Fulton Books 2022

ISBN 979-8-88505-423-2 (paperback)
ISBN 979-8-88505-424-9 (digital)

Printed in the United States of America

Contents

We Will Rise

If a stranger comes to our house, we should offer them a plate.
And it doesn't matter if they've already ate.

If a person is hungry or not, who are we to judge.
We should know that it's our Christian duty to always show love.

We go thru heartaches and tragedies in our life.
And sometimes just getting out of bed can turn into a fight.

If we turn our hearts over to Jesus, He'll make heaven our prize.
And no matter what we're going thru, we will rise.

Didn't Jesus feed thousands with little to nothing.
Proving that when it comes to giving, we all have something.

That's why God put people in our midst to help guide us straight.
Someone who knows and believes that only love can conquer hate.

Someone to help us lift the blinders from our eyes.
Constantly reminding us that no matter what we're going thru, we
    will rise.

We must find it in our hearts to forgive those who have done us wrong.
And let the God inside of us keep us strong.

God knows that some people are just stronger than others.
And that's why He put us all here to carry one another.

Remember we will be judged by the things we say and do to others.
How we personally cause someone's life to have peace or struggles.

It doesn't matter if we're going thru a pandemic or being persecuted
    by lies.
As long as we have faith in God, we will rise.

Dear Mother

Dear mother, with loving eyes and voice as sweet as an angel.
Thank you for never giving up on me and for protecting me from danger.

No matter what I do wrong, you're always there encouraging me to
    do better.
Even though I let you down from time to time, you're there for me
    during stormy weather.

When I am in a bad relationship and heartbroken with pain.
You always know how to keep me sane.

Every time I've needed to feel love, I can always depend on you.
You're always by my side, no matter what I'm going thru.

I could never give back what you've given me because of you I can
    truly say.
That the morals and values you've instilled in me have made me the
    person that I am today.

I know what love is because that's all you've shown me since the day
    of my birth.
You mean more to me than anything in this world; there's no price
    to your worth

Thank you for not giving up on me when everyone else turned their
    backs.
And out of all the unions that exist, there's none stronger than a
    mother and child pact.

The only love that comes close to the love of God is the love of a
    mother.
And I thank God every day that you are mine because you can't be
    replaced by no other.

I'd Rather Have a Peace of Mind

I'd rather have a peace of mind than all the money in the world.

To me a peace of mind is worth more than diamonds and pearls.

Look at King Saul, with all his power, he was filled with a troubled spirit.

Because of his disobedience to God, there was confusion in everything that he did.

When we turn our lives over to Jesus, He'll shield us from any harm.

We may not have everything that we want, but we'll have peace amidst the storms

Jesus said that it is easier for a camel to go thru the eye of a needle, than for a rich man to enter into heaven.

That's because of the distraction that money causes in this world that we live in.

They say that money can buy love, it can even buy respect.

Until we run out of it, and we're left with nothing but regret.

But when you're broke and all alone because everyone turned their backs.

I want you to know that Jesus will always be there to take up the slack.

In everything that happens in our life, we're going to face some kind of test.

After all, that's how God brings out our best.

I'd rather live right and do without in this world that we live in.

If that's going to get me a place in heaven.

God frowns every time He sees what we are willing to do for a dollar bill.

Some even say that what they are doing is a part of God's will.

If we think that money is the key to all our problems, there's something wrong with our heart.
Remember it says in the book of Proverbs, "A fool and his money will soon part."

If what we're doing to get money is not God's way, we need to change our way of thought.
Because we're in for a big letdown if we think that a ticket to heaven can be bought.

The only way we're going to find peace is through Jesus Christ.
And until we do that nothing that we do will be right.
Only through faith will we have the strength to believe.
That nothing that this world has to offer will ever give us peace.

Remember those who are last on earth will be first up above.
There is no better illustration of the goodness of God's love.

So when we see ourselves walking around with a troubled mind.

We have to keep our faith in Jesus because He will always be on time.

Never Give Up

I don't care who we are, we'll never know God's plan.
Don't the Bible tell us to lean not to what we understand?

Yet there are people out there predicting other people's life.
Preying on those who are separated from the light.

Using the Bible for their personal gain.
Running people away from the church without an ounce of shame.

Thinking that they can do no wrong.
Preying on the weak, yet running from the strong.

You see the enemy can come in all shapes and forms.
With just one purpose in life, to cause us harm.

And that's why we have to shake that weight off of our shoulder.
And pray to God to make us stronger.

Get out of the darkness and come into the light.
So that God almighty can give us peace in our life.

He can do it, we just got to believe.
It's the devil, not God who likes to deceive.

With God by our side, we can climb any hill.
It's not what man says, it's all about God's will.

After all He is the one who created us.
He's the one who has unconditional love.

He's the one who created the moon and the stars.
He was the one who sent us our Lord.

He can take all of our enemies and place them under our feet.
And give us strength during those moments when we are weak.

The best way to beat hate is to shower it with love.
And have faith that through God, all things are possible.

We are always getting caught up in a spiritual fight.
Being mentally attacked during the morning, noon, and night.

The devil's working overtime, but He doesn't work more than God.
And that's why He's always being defeated by God's staff, by God's
    rod.

It's time to fight the lust and embrace God's power.
And stop letting downfalls in our life make us sour.

The greatest key to success is dedication.
But it would have no purpose without motivation.

Why do you think God gave us one mouth, but two ears.
So that we can get a better understanding of the things that we hear.

We'll never learn anything if we refuse to listen.
And that's why a lot of us are constantly missing out on our blessings.

Some people are just addicted to being seen.
They ignore the red light because all they want to see is green.

In order to be a good leader, we've had to be a good follower too.
That's the only way to learn what a good leader must do.

We have to turn our lives over to Jesus and let Him fill our cup.
And no matter what we are going through, we can never give up.

You Can't Make Me Cry

The hate that you feel for my skin color, I can never understand why.
But no matter what you do to me, you can't make me cry.

Anytime you see me doing good, it burns you up inside.
You can love me, you can trust me, all you have to do is try.

Reaching out to you desperately, but no matter how hard I try.
I'm left with so much disappointment because I just can't understand
why.

I pray every day that I could just live in a normal world.
Where hatred and prejudice didn't exist because everybody cared.

What is it that intimidates you about the color of my skin?
How can you judge a book by its cover without attempting to read
what's within.

When you see my skin color, you see an animal, I just want to know
why.
What kind of a person feels good inside, when they see another person cry?

When I see the hatred in your eyes it's way too hot for me to touch.
I just want to be treated equally; I'm not asking for much.

I am so tired of some people thinking that being black is a sin.
And why am I being treated like a criminal, just because of the color
of my skin.

Injustice, poverty, and racial discrimination is bad for us all.
And it's no one but the devil encouraging others to want to see a race
of people fall.

Emancipation proclamation sounds good written on paper.
But what good is it, when a race of people are being extorted for their labor.

I've been spit on, lied on, called the N word, and economically deprived.
But no matter what it is that you've done to me, you still can't make me cry.

It doesn't matter what religion, what race or where we come from.
As long as we have Jesus in our life, we will overcome.

Pressure is a privilege, that's the only way that we become strong.
WE can conquer hate with love, but we can't do it alone.

Everything that you've done to me and my ancestors should have caused malice in my heart.
But I know that I'll never make it to heaven, if I let hatred tear me apart.

Sometimes we have to step out of who people are and realize what they are meant to be.
That's the only way that we'll ever be ready, to set our souls free.

There is an ongoing war in a person's heart; one is love, the other is hate.
And it depends on which one we choose to follow that's going to decide our fate

Not a person who's ever lived has not had a bad moment in their story
Which is why we all have a testament of God's mercy, grace, and His glory.

It's never too late to change how we think, the hate that we feel.
God has already instilled inside of us, a love that's positively real.

The second greatest commandment from God is to love others as we love ourselves.

It says nothing about us being better than someone else.

And for all of those people who refuse to accept the truth.

I don't care how much you hate me, how much you hurt me, I still have much love for you.

Father Said There Would Be Days Like This

When we have nothing of value in this world to call our own.
And we've been kicked out into the streets and left without a home.
And the rain is constantly pouring down on our life.
And just getting out of bed is an everyday fight.
When we are having issues inside of our house.
Because we are having marital problems with our spouse.
And our children are getting into trouble because they've got caught
    in the street.
And we are living pay check to paycheck week to week.
Father said there would be days like this.
He told us not to give up, He told us not to quit.

When we are looking for a change, but we are still seeing the same
    faces.
Because we are searching for love in the wrong places.
And we're feeling so lonely, our heart is aching with pain.
And we're doing everything in our power to keep from going insane.
When we're having a hard time paying our bills.
And no one can seem to understand just how we feel.
When our belly is aching because we have no food.
And we're being mentally or physically abused.
Father said there would be days like this,
He told us not to give up, He told us not to quit.

When races are against races and nations are against nations.
And prophecies are being fulfilled that are written in Revelations.
When mothers and daughters are at war with one another.
And just to have peace in our lives seem like an everyday struggle.
When we lose someone to death that we loved so much.
And every single day we yearn for their touch.
When everyone seems like their losing their minds.
And no one is paying attention to God's heavenly signs.

11

Father said there would be days like this.
He told us not to give up; He told us not to quit.

He said that there would be nowhere for us to hide.
He also said that He would never leave our side.
The Bible tells us that we should not lean to our own understanding.
We have to be patient and stop being so demanding.
No matter what we're going through someone else is in worse shape.
After all, we are living in a world that's filled with hate.
A world where people are being judged by the color of their skin.
A world where people don't think that hate is a sin.
Father said there would be days like this.
He told us not to give up; He told us not to quit.

Now when everything is going new—good our way.
And we are no longer having those stressful days.
When we are feeling nothing but peace when we walk thru our home.
And we are no longer feeling sad and all alone.
When our bills are paid, and we are feeling no stress.
Because we put our faith in God, even though we have less.
This is why we couldn't give up, why we couldn't quit.
Because Father also said there would be days like this.

Everyday Beggar

I don't know about anyone else, but one day I'm going to walk
    through those pearly gates.
But I'll never see heaven if I allow myself to be controlled by hate.

We have to stop telling others where their souls will end.
God didn't give none of us the power to judge another person's sin.

It doesn't matter what color we are, or what home we come from.
When God sees us, He sees us as His daughter or as His son.

We could have all the money in the world or have the strongest mind.
But none of that will matter when God says that it's our time.

So the next time we put ourselves up on that pedestal, and think
    we're all that.
We better take a good look in the mirror and accept the true facts.

We don't have anything if we don't have God in our life.
And no matter what we're doing, without Him it won't be right.

WE have to have faith that there's no problem that He can't solve.
Why do we worry, if we really believe in God.

Hasn't He always been there for us when we needed Him the most.
And don't His love reach out far and wide, from coast to coast.

Oh wasn't that a great feeling when we accepted Jesus in our heart.
How He forgave us for our past sins and gave us a brand-new start.

We think that we are in control, but our lives have never been ours
    to make.
And that means that a life wasn't for us to take.

If it wasn't for God, none of us would even have life.
He's the one who's watching over us morning noon and night.

But instead we walk around acting like we are better.
Just because God has brought us through our stormy weather.

We talk about people when we see them begging on the street.
And make fun of them by calling them weak.

But when it comes down to it we're no better than they are.
When it comes to God we are all everyday beggars

I don't know about you, but I ask God for something every single
day.
As a matter of fact, when I am in need the first thing I do is pray.
God has shown me over and over again.
That I need His help if I am ever going to be delivered from sin.

We can never be so proud that we can't ask for God's help.
And we can never have so much, that we think we're better than
someone else.
There is no limit to what God can do in our life.
If we just repent for our sins and believe in Jesus Christ.
We can't listen to those people who say that God doesn't exist.
As long we believe, we'll be spiritually rich.

But if we want our names written in God's Holy Book.
We better stop judging others just because of the way they look.

How can we think that we're better than anyone else.
When we're nothing but everyday beggars ourselves.

Soul Reason for Living

When I was sitting all alone and misguided in the dark.
You lifted up my spirit and restored love back into my heart.

As the tears flowed from my eyes because of the destruction in my
life.
You are my comforter; you took away my sleepless nights.

I was stripped of all my self-esteem and honor no longer existed.
As I wondered blindly through this world with my days and nights
twisted.

But no matter how low I felt, you never allowed me to give up.
And no matter what I was going thru, you never overfilled my cup.

To this day my Lord, you are my soul's reason for living.
You taught me how to love others, and the art of giving.

When I think back to the things I've been thru, it brings tears to my
eyes.
I was the reason for all of my problems, but I couldn't understand
why.

Until you came into my life, I had no reason to live.
I was very good at taking, until you taught me how to give.

I didn't know how to love because I didn't love myself.
When you came into my life, I was one step away from death.

To this day my Lord, you are my soul's reason for living.
You taught me how to love others, and the art of giving.

Now I can be a better father to my children and a better husband.
If you hadn't come into my life, my heart would be filled with grudging.

15

Just maybe I can be the kind of man who makes their parents proud.
As I testify to your goodness and mercy, speaking clear and loud.

Where I was once lost, I'm found, you kept me out of my grave.
My whole life was changed for the better, since that day that I was
    saved.

To this day my Lord, you are my soul's reason for living.
You taught me how to love others, and the art of giving.

Just Another Learning Lesson

When you're in the midst of a storm, keep your head up even if it
    may seem too hard.
And stay focused because there is a task that God is leading you toward.

We're not smart enough to understand what's happening at God's level.
Most of the time we're led astray by listening to the devil.

We are all put here to serve God's purpose.
Which means that none of us are worthless.

First we must realize that no one is perfect, we all make mistakes.
That's why we have to find our way to Jesus no matter what it takes.

And if we don't put God first in our life, we're in for rough times.
Without Him in our life, it's impossible to have a peace of mind.

So don't let the devil fool you into thinking you're all alone.
He's just trying to keep you from reaching your new home.

Jesus said, "Up in my Father's house there are plenty of rooms"
Jesus is up there preparing a place for us now, and He'll be returning soon.

When you feel like you're at the end of your rope, and you can't take
    no more.
Then it's time for you to call on Jesus and invite Him thru your door.

Then you'll see that all of your problems will all seem small.
Jesus is ready to answer your prayers; He's just waiting on you to call.

Just keep your faith and you'll receive your blessings.

Remember what you're going thru isn't the end of the world; it's just
another learning lesson.

That Could Have Been You

How do I know that God is inside of me, I can feel His presence in
    everything I do.
And it feels good to finally open my eyes up and accept the truth.

With nothing to hide and no reason to be ashamed.
Treating my life no longer like it was some kind of game.

But my life didn't just get this way overnight.
I had to be knocked down and brought to my lowest point in life.

During those times when I was going thru things that I couldn't
    understand.
God was transforming me into a better man.

Before I thought I could solve all my problems by myself.
To blind and full of pride to realize that I needed God's help.

Even as I was falling Satan prevented me from reaching out for some-
    one's hand.
I was slowly drowning in a pit of quicksand.

Suffocating, while the very breath was being squeezed from my soul.
Being condemned by my own heart for the very things that I couldn't
    control.

But right when I was about to die and take my last breath.
I called out in desperation to Jesus and immediately He came to help.

Then a feeling came over me that I just couldn't explain.
I was amazed and astonished by the power in His name.

As He lifted me up and set me back on my feet.
Tears of joy stream down my face, as I began to weep.

God had a revelation for my eyes that He wanted to show.
He told me that there were a lot of people out there just like me who
    needed to know.

That it is a wonderful feeling when you have God in your life.
He's the only one who can bring us out of darkness and into the light.

Not a day goes by that I don't thank Him for giving me a choice.
There was nothing more beautiful than to hear His voice.

Oh, what a joy it is to have peace in my life.
And I owe it all to my Lord and Savior, Jesus Christ.

So now when I see someone homeless or strung out on alcohol and
    drugs.
I am no longer afraid to walk over and show that person some love.

And when I see someone that I know who's lost in the streets.
I thank God for His unchangeable love because that person use to
    be me.

So when we see that mother, that daughter selling her body for drugs.
She needs encouragement and understanding, not to be judged.

We never know what they're going thru until we've walked in their
    shoes.
And if it wasn't for God's presence in our lives, that could easily have
    been me or you.

So instead of lifting our noses to others and judging someone we
    don't know.
If we pray for them and help out the best way we can then our bless-
    ings will flow.

We were all sinners at one time in our life, this we can't deny.
And we would still be out there, if Jesus hadn't answered our cry.
So the next time you see that dirty man on the corner asking for
money don't be rude.
Just remember if it wasn't for God's presence in our life, that could've
been me or you.

My Last Goodbye

When my time comes I want my family and friends to rejoice and
   sing.
And keep in mind that when you're a true believer dying is a wonderful thing.

I'm ready for my homecoming, which I know will turn into a reunion.
walking hand and hand with Jesus on the other side of the sun.

I am reflecting on that day when I turned my life over to Christ.
When He came into my heart and changed my whole life.
He made a difference to me, and He can do the same for you.
Just turn your life over to Jesus, and He will see you thru.

Just remember God's promise that He'll never leave your side.
Which is why I'm cheerfully giving you my very last goodbye.

Goodbye dear sister, whose strength I've leaned upon.
Shed no tears of sorry, on this day my soul has won.

Goodbye, my dear brother, you've protected me throughout my life.
Let not your heart be troubled, I'm standing next to Christ.

Goodbye, my dear children, I know your hearts are filled with pain.
When you find yourself in need, just call on Jesus's name.

Goodbye, my dear wife, I know that God put us together.
And even though I am no longer there with you, our love will last
   forever.
When you think of me don't cry, rejoice, be happy and just laugh.
And when you make it to heaven we will cross each other's path.

Goodbye, my dear parents, you were an inspiration in my life.
You taught me how to be a good person by always teaching me right.
I know that you're going to miss me, but dying's a part of life.
Remember that's the only way to come face to face with Christ.

And last but not least goodbye my dear friends, you've been there
    through thick and thin.
You've given me the support that I've needed, to the very end.

To the rest of my family, I now say my last goodbye.
If you ever find yourself afraid and lost, let Jesus be your guide.

No longer am I suffering, no longer am I in a strain.
I'm going to a place where I'll no longer feel no pain.

So please don't cry for me, unless their tears of joy.
I'm in a better place than I was, standing next to my Lord.

Don't Lose Your Faith

Death is not the greatest loss in life, the greatest loss is what dies
　　inside us while we live day to day.
Our trust, our self-esteem, our love, even our faith.

True believers don't fear death because we know that dying is a won-
　　derful thing.
We know that that's the only way to come face to face with the King
　　of all Kings.

The Bible tells us that if we have enough faith, we can make moun-
　　tains move.
It also tells us that faith is one of our greatest tools.
It's the reason why we love a God whom we've never seen.
It's the reason why we believe that our souls have been redeemed.

We all go through bad situations in our lives that cause us pain.
So why do we choose to judge others just because their pains are not
　　the same

We have to understand that none of us is perfect, we all make mistakes.
And the day that we look at ourselves in the mirror and lie to our-
　　selves that's the day that we become fake.

That's why we have to stay prayed up from day to day.
And if we don't put love in our heart, we'll be consumed by hate.

God wants us to love our sisters and brothers as much as we love ourselves.
But if we don't love ourselves first, how can we love anyone else.

No matter what we're going through our pains are nothing compared
　　to what Jesus faced.
And that's why we have to thank God every day for His love, mercy,
　　and His Grace.

The gift of heaven has always been meant for all of us to reach.
So if we want to send out a message from God, then faith is what we
    should teach.

None of us know when it may be our time to go away.
WE never know when certain words would be the last thing that we
    say.

That's why whatever we're doing, we have to have God in our life.
He's going to always love us, and it doesn't matter if we're living
    wrong or right.

That's why we have to stop worrying and listening to someone else.
If God hasn't given up on us, then why are we giving up on ourselves.

We were all born with our very own special gift.
Something that we are especially good at, something that gives our
    lives a lift

We can't. let the enemy get into our mind and twist the truth around.
He'll have us waking up every day feeling like our souls are in the lost
    and found.

We can't let him keep turning us against our sisters and brothers.
WE need to have more compassion for another person's struggles.

We can't keep letting him shatter our dreams.
With God in our lives we can overcome anything.
For God so loved the world that He sent His only begotten son.
And He that believes in Jesus has already won.
We all have the power to put the enemy in His place.
As long as we don't let anything or anyone make us lose our faith.

Turn the Other Cheek

Being a Christian is about what we do and who we know.
Which is why we should associate ourselves with those who can help
us spiritually grow.

Why would we want to be around someone who always put us down.
Someone who loves to see us frown.

I remember when I was a young boy back in school.
How the kids would be nasty with the insults that they used.

They would talk about your mother, your father, your sisters, and
your brothers.
They did it for a show because they did it in front of others.

All the other kids wanted to see a fight.
But because you were a Christian, you knew that wasn't right.

You knew that Jesus said to turn the other cheek.
But some of those kids made you want to knock them into next
week.

Just because we turn the other cheek.
That doesn't mean that we are weak.

Anytime someone comes at you with hatred in their heart.
Kill them with kindness and then introduce them to the Lord.

There are many battles that we can't fight alone.
Which is why we can't afford to have a divided home.

Everyone in the home play a big role.
When it comes to making heaven our number one goal.

We have to pray that God delivers us from our sins.
By first ridding us of the hatred within.

Who really knows what they'll do when put to the test.
Will it pull us down or will it bring out our best.

Each one of us is gifted with amazing strengths.
Straight from God, heavenly sent.

So just because we turn the other cheek.
Don't let anyone tell you that you are weak.

All of us have said things before that we didn't mean to say.
And some of us are still regretting those words right to this day.

So what's so hard about admitting when you are wrong.
Especially when it can restore peace within your home.

Somebody has to play the role of the peacekeeper.
And it shouldn't be hard if you are a true believer.

Let's keep the policy of the open door.
And start listening to our children more.

Remember just because we turn the other cheek.
Doesn't mean that we are weak.
And turning the other cheek ain't so hard.
As long as we're turning it for the Lord.

Harmony

Music is a way of communication for people in the streets.

They say that music is what it takes to tame the beast.

Just a simple melody will make us laugh or it'll make us cry.
Some words are the difference between wanting to live or die.

The heart is capable of holding back so much pain.
And there's no better feeling than singing to God in the rain.

We sing because we're happy, we sing because we're sad.
We sing when we are good, we sing when we are bad.

Even when we are afraid of singing in front of someone else.
We are never too shy, to sing for ourself.

Our voices seem to always have more power.
When we are all alone, singing in the shower.

Even though they say that everybody can't hold a note.
Don't let anyone stop you from singing for the Holy Ghost.

After all it was God who blessed us with a voice.
And whether or not we choose to use it, it was God who gave us that
    choice.

We were all born with a special gift.
Something to give our lives a lift.

No matter what we're going thru in our life, Jesus is the key.
And if God put singing in our heart, then let's sing with harmony.

If God Can Forgive then so Can You

We say we love God, whom we never laid eyes on.
Yet we can't stand the people living in our own home.

Just because someone said or did something to make us mad.
We lose respect and understanding in an instant flash.

Some of us can even hold a grudge until the day we die.
Just because someone made us cry.

We want to be forgiven for the wrong that we've done.
Yet we continue to abuse others with our tongue.

Misery loves company and the admission is free.
And faith is nothing if we don't first believe.

Some of us even bring this attitude to the church.
Not wanting to forgive because we were hurt.

Most of the time we don't want to accept the truth.
If God can forgive, then so can you.

Love is a word that is so easily said.
But if we don't have forgiveness in our hearts, then our love is dead.

So if you'll hold a grudge, regardless of the reason.
Find a way to forgive and this may be your season.

Meditate your mind, talk to God and no one else.
And gain peace in your heart, by first loving yourself.

Accept Jesus in your life and accept the truth.
If God can forgive, then so can you.

We're All God's Kids

What makes you think you're better than me, is it the clothes I wear.

Every time you come around all you do is stare.
I may not have a job, and I often go without food.
I know how you feel about me; I'm no fool.

I do believe that if you look hard enough you'll find good in everyone
you meet.

How do you know if that person does or doesn't have goals they're
trying to reach.

I believe that when you find fault in another, it's a reflection of how
you see yourself.

And you have issues in your life that are quietly kept.

The only way to make you feel good about what you are doing.
Is to target someone's life with the new—intent to ruin.

You can't see that all your problems start from within.
It's time that we all realize that a sin is a sin.

So take a look in the mirror, and see what comes to your mind.

Stop judging others and leave the past behind.

If you have a problem with another, let it be for something they've
done.

And stop thinking that you're better than others, we're all God's
daughters and sons.

He Rose on that Third Day

Jesus was labeled as a sacrificial lamb before He was even born.
Accused of serving the devil and wearing horns.

Even though He spent His whole life preaching nothing but peace.
The heads of the temples cared for Him the least.

To this day there are many who refuse to admit that He is the way.
But it was Jesus, not Buddha nor Muhammad, who died and rose in
    three days.

Not only did He prove that He could overcome death.
He also went down and took the keys away from hell.

They whipped Him and made Him wear a crown of thorns, as He
    was forced to carry His cross.

But just because He allowed these things to happen, doesn't mean
    He was soft

It was His purpose to die so all of our sins would be taken away.
And it was Jesus, not Buddha nor Muhammad, who rose from His
    grave on that third day.

Not only did He come back from death, He also walked throughout
    the land.
Just to show that there was no limit to the love that He has for man.

It was His sole purpose that we all be given a chance.
As He preached to all who listened, that the Kingdom of heaven is
    at hand.

So while we're teaching our children about easter eggs on this glorious holiday.

Just remember that it was Jesus who died on that cross for us, and arose from death on that third day.

Imitations of Life

Who am I, and why was I even born.
Am I special or am I just another man being formed.
Sometimes I wondered if I would be more important in this society.
If being born into a rich family was a reality.

Would I choose to be black or would I choose to be white.
Why am I so caught up in these imitations of life.
Who wouldn't want to be born with a silver spoon in their mouth.
Not ever having to worry about growing up in a rat-infested house.

I don't care how poor we were, I am blessed just to be me.
Especially since I come from a loving family.

I thank God every day for allowing me to be myself.
Because if I don't love me first, how can I love someone else.
And as long as I continue to follow Jesus Christ.

I don't have to deal with these imitations of life.

Oh, what a joy it is to have Jesus as my friend.
Cradled and protected by His heavenly hands.

If You Think It's Hot Now, Keep on Sinning

If you think it's hot now, keep on sinning.
These 100-degrees-plus days are only the beginning.

There will come a day when your thirst can't be quenched,
and that's because living right was what you stood against.

Don't let the devil fool you into thinking hell doesn't exist.
Deceit and deceptions are his number one tricks.

He wants you to believe that you don't have to
follow God's rules to get to heaven.
He wants you to believe that salvation is
hard to get even though it's given.

His days are numbered, and there's nothing that he can do,
except try to convince mankind that what
is spoken in the Bible isn't true.

He wants to turn sons against fathers and
daughters against mothers,
it suits him well to see us at war with one another.

We're nothing to him, just toys that he plays with,
so why do we condemn ourselves for sinning
instead of accepting God's gift.

It's going to be a sad moment on judgment day
for those who haven't accepted Christ.
Their soul will be condemned to the fiery pits
because of how they lived their life.

So if you haven't accepted Jesus as your Lord
and Savior, this may be your last chance.
No longer can we afford to play both sides;
the time of judgment is at hand.

God loves us so much that He sacrificed His son.
Who died on the cross so our souls could be won.

And He doesn't expect us to be perfect at all,
but He does want us to obey Him when He calls.

Believe me, hell is one place that you don't want to go.
But if you keep on sinning you will condemn your soul.

God has a plan and that's for all of us to be saved.
Which was made simple by the price that Jesus paid.

Don't let Satan fool you into believing you can't be saved.
Repent for your sins, and give your life to Jesus this day.
But if you like the heat, then keep on sinning.
These 100-degree-plus days are only the beginning.

It's Never too Late

Once the drugs and alcohol are gone, there's an empty void to fill.
But this too can be accomplished as long as we possess the will.

Those bridges that were burned down a long time ago.
Can be restored if you have the desire to see your future grow.

But first we must deal with life after the drugs and alcohol are gone.
And realize that just because we're sober doesn't guarantee
that we'll have a peace in our homes.

The very things that were avoided for so many years.
Will become a reality that could cause a flow of tears.

The ones that we love will still have a problem with trust.
With what we put them through, we have
to know that this feeling is just.

The many flaws that we have, which we
tried to cover up with drugs.
As we look in the mirror each day will cause us the shrug.

But we have to keep hope in our hearts each and every day.
And realize that through God that it's never too late.

Yes, those old temptations will often arise.
And Satan will try to fill our heads up with lies.

But as long as we're seeking better tomorrows.
With Jesus in our lives we'll be able to overcome all of our sorrows.

God has a purpose for each one of us.
And He demonstrates this every day through His love.

So if a person doesn't trust you, that doesn't
mean that there's no love.
Because trust is the hardest thing to restore
when it comes to people like us.

Yet, let your actions do your talking, and keep your head lifted high.
Because an overzealous tongue can produce nothing but lies.

Just tell yourself you gone make it day after day.
And remember when it comes to restoring
your life, that it's never too late.

What If Jesus Came Today

What are you going to do if Jesus came today?
Are you going home with Him or will you stay?

If you have to think about this question twice,
then you need Christ in your life.

It's going to be a sad day for those who don't believe.
They're going to be left behind no matter how much they plead.

What are you going to do if Jesus came today.
For those who didn't accept Him it's going to be too late.

He's sent many messengers for us all.
To change our ways before we fall.

He's given us many chances over the years.
But the day has come for a sinner's biggest fears.

When He comes we'll all be judged.
By the goodness of His love.

So what are you going to do if Jesus came today?
If you have any doubts, then it's time to get saved.

God Is Real

When we are going thru trials and tribulations in our life.

We have to keep our faith and believe that God is always right.
He may not come when we want Him to.
And some of us even blame Him for the bad things that we're going
through.
Whether we're experiencing poverty, sickness, or the death of some-
one we love.
Whatever we're going thru God will never abandon us.

The ones who panic and give up in this life.
Are the ones who don't know Jesus Christ.

We can overcome any obstacle that we face.
If we believe in God, if we have faith.

Problems are a part of life; everything's according to God's will.
But we can't understand that, if we don't believe that God is real.

And if we do believe that God is real, then why do we worry.
When we should stand firm on His word and always give Him the
glory.

We've all suffered mentally or physically sometime in our life.
And it had nothing to do with living wrong or right.

We as believers have to always trust in God and keep the faith.
And know in our hearts that only love can conquer hate.

People are in the stores fighting over supplies.
Doing whatever they need to do just to survive.
They're more afraid of a virus than they are losing their soul.

When they should be making going to heaven their goal.

Sickness is a part of life; it's in accordance with God's will.
But you'll never understand that if you don't believe that God is real.

We can't let what's going on in this World stop us from living our life.
We have to stay prayed up and continue to do what's right.

People die every day, and until Jesus comes back, that'll never stop.
That's why our feet have to be planted on solid rock.

We have to remember that the devil can take on many roles.
And that's why we can't let nothing or no one make us lose our souls.

Death is just a part of life, and it's in accordance with God's will.
That's why we have to let everybody know that God is real.

And that's why we have to stay strong and show the world God's true
      light.
But we can't do that if we don't introduce them to Christ.

We have to let the non-believers know where Jesus brought us from.
About all of the miracles He has given us since the day that we were
      born.

This is not the time for God's people to run and hide.
We can overcome anything as long as we have God by our side.

We need Jesus in our life to beat the devil.
No matter how many tricks he have, he's not on God's level.

Destruction is a part of life in accordance with God's will.
We need to let the whole world know that He is real.

I'm going to continue to thank God each and every day.
If it wasn't for Him, I would've never been saved.

And no matter what's going on in this world, we have to show love
and respect.
And believe in our hearts that God always knows best.

At the Age of Twelve

Why do I want to be a knucklehead; there's no benefit, I get nothing
    in return.
All I do is keep making mistakes, am I gonna ever learn.

I'm not too sure if my friends are real, but I tell myself that they're
    cool.
And every time one of them turn their back on me, I feel like a fool.

I even tell myself, that I am the head of the pack.
And if anything happens to one of my friends, I always got their
    back.

All that thinking did was land me behind prison walls.
Who would've ever thought that at a young age, someone like me
    could fall.

On the street corners rapping, giving honors to a fake savior.
Not realizing that all of this was nothing but bad behavior.

Then I read about this kid who was preaching to grownups at the
    age of twelve.
He was telling these grownups about heaven and how to avoid hell.

It was then that I realize that you're never too young to make a
    difference.
But first I had to experience my very first deliverance.

Now I'm no longer living behind those prison bars.
Fighting my insanity because of my spiritual wars.

All I had to say was, "Jesus, take control over my heart.
Forgive me for my past sins, please give me a new start."

Then all of my problems seem to fade away.
And the peer pressures that I once had no longer bother my days.

And it's all because of that child who was preaching to grownups at
the age of twelve.
Telling them if they seek the kingdom of heaven, they would need
nothing else.

God Already Knows Our Story

What good is it to gain this world, but lose our soul.
When getting to heaven should be our number one goal.

We can't keep putting our spouse, our children, or our friends before
    Christ
After all it was Jesus who died on the cross so that we could have
    everlasting life.

It was always meant for heaven to be all of our home.
But only through Jesus can we reach God's heavenly throne.

Jesus came here to save, not judge the world.
He wanted all of us to know just how much God cared.

God already knows about our pains; He knows about our fears.
He's just waiting for us to call on Him, so He can wash away our tears.

That's why we can't put nothing or no one above Him, only through
    Him is there salvation.
After all, He's the whole reason for our creation.

And why would we want to go to heaven, if we have everything we
    want on this earth.
I don't care who you are or what you have, we all go through some
    kind of hurt.

The best thing that we can do for our loved ones is pray for their
    souls.
And ask God to put it in their hearts to make heaven their number
    one goal

If we encourage one another I believe we can make this world better.
But first we have to get along and learn how to work together.

It's time for all of us to start believing in miracles.
Whatever we're going through, we have to believe that God is a
   deliverer.

God doesn't force us to do anything against our will.
And even though we've never seen Him, we have to believe that He's
   real.

He will whisper in our ears and offer us suggestions.
And even when we're doing wrong, He'll shower us with blessings.

But it's up to us whether or not we are obedient.
We have to have faith that Jesus is the main ingredient.

He was the one who died on that cross on top of calvary.
And He didn't do it for Himself, He did it for you and me.

So that our sins could be forgiven, and our souls could go to heaven.
And there was nothing that we did to deserve it; our salvation was
   given.

And let's stop giving credit to the devil.
I don't care how conniving he is; he'll never be on God's level.

The very things that we love the most can be taken right out of our
   life.
Especially if we're showing them all our love and not giving any to
   Christ

It's sad when we worry about everybody else's soul but not our own.
And they end up in heaven, and our souls end up at a different home.

When it comes to going to heaven, we should all have that thirst.
But we will never get a chance to see it if we don't put God first.

We have to stop attacking other people's faults and take a self-inventory.

We might fool the people around us, but God already knows our story.

You Are Special Because I Made You

How can you still love me after all the wrong that I've done.
I've caused so much hurt and pain to some.

I've taken credit for all the blessings that you've given to me,
and I can't remember the last time that I got down on my knees.

As far as I know I've been selfish all of my life,
and my way was always considered in my mind to be right.

I didn't care how I treated my neighbors,
and I cursed out everybody looking for help at my doors.

I'm even cruel to animal as I see them on
the streets. Nothing that I come
into contact is safe from me. But no matter
what I do, you still seem to care.
Tell me what's going on; please make me aware.

How can you love someone like me God, what is your take?
He said, "You are special because I made you,
and I don't make any mistakes."

"I've been dealing with sin, from the moment
I threw Satan out of heaven,
and when it comes to patience, there's no one at my level."

"So keep doing what you are doing, believe me, you'll get tired.
Which is the same thing I told David
before he broke down and cried."

"There is nothing that you can do that I haven't seen before,
but when I get through with you, you won't sin no more."

Changed

I know I caused you pain. I know I caused you sorrow.
If you believe that Christ has changed my life, we
can look forward to a better tomorrow.

I'm not making no excuses for the misery I've
caused you during my addiction.
I can't stand the thought of losing you, so I'm
making this my number one mission.

Repair my life, stand tall like a man, and
start doing what is pleasing to you.
Going back to those days when we stood as one at the altar,
as we both spoke those words, "I do."

I know that my addiction took away many years
of our lives that can never be replaced.
and if you'll never forgive me and move on
with your life, this I'll just have to face.
I just want you to know, there'll never be
another who could ever take your place.

Father, I Thank You

Father, I thank you for delivering me from myself and introducing
    me to your light.
You are worthy to be praised morning, noon, and night.

Your love has brought me from the thresholds of death many, many
    times.
And without your words of comfort. I would have easily lost my
    mind.

You've lifted the covers from my eyes so that I could see.
I know that whatever I am going thru, you are always there for me.

There is no greater power than that that you possess.
And I know that with you in my life, I am truly blessed.

I thank you, Father, for delivering me from the wickedness of the
    streets.
I know that it was you, who was constantly looking after and pro-
    tecting me.

It was because of you, Father, that my life was spared.
You never gave up on me, even when no one else cared.

I thank you, Father, for watching over my children.
And for providing them with the things that they need, in this world
    that we live in.

I thank you, Father, for giving me the will to get down on my knees.
I know that without you I would not be free.

I thank you for your son, Jesus, the one that you have sent.
And thank you for your forgiveness every time I repent.

If it wasn't for your love I would've died a long time ago.
Because of the terrible things I've done in my life, that only you and
I know.

There hasn't been one time when I called, that you didn't come to
my aid.
And there is nothing that can compare, to the price that you've
already paid.

That's why I am standing here today, with tears of joy in my eyes.
To thank you, Father, for coming into my life, and causing me to
rise.

Love Conquers All

I've let many people down in my addiction, that's why I now try so
hard to do my part.
And if it's hard for anyone to forgive me, I'll hold no bitterness in
my heart.

It's amazing how one action can cause a chain of events, that could
alter our lives.
And I thank God for finally opening up my eyes.

Drugs can take away the very essence of good that is embedded inside
of us.
And fill us up with so much hate, that it can only be conquered by
love.

I never knew how others felt until I myself became the victim.
My life swung so far out of control that I became a part of the penal
system.

As I was locked up behind bars, I was told to praise and rejoice.
That was when I first heard God's voice.

God told me that in every storm that I go thru, there would be a
rainbow at the end
He also told me that the love that He has for me would never break
nor bend.

It doesn't matter what wrong is done to us; we should always keep
love in our heart.
And once we accept Jesus as our savior, all of our blessings will start.

Through love there is hope that our future will be bright.
But the only way to find real love is to turn our hearts over to Christ.

So when we find ourselves having a difficult time forgiving others.
We have to remember that thru God we are all sisters and brothers.

No matter what we do He always forgives us.
God is all about forgiveness and love.

So if He can forgive me for the many wrong things that I do.
It would be wrong for me to hold any malice in my heart for you.

And as of today I see you for where you're going not where you've been.
Because of the love in our hearts we can once again be friends.

It doesn't matter what color we are, we can be fat, thin, short, or tall.
We'll be forever blessed as long as we realize that love conquers all.

They Said

They said that I'd never amount to nothing; I'd always be an addict.
They said that I wasn't worth saving because I wasn't their pick.

They said that I couldn't think for myself because
I was making the wrong choices.
They said that I was irresponsible because
I wasn't listening to their voices.

What they don't know is that God had hardened
my heart for a purpose He had in store.
All my blessings were constantly adding up;
they were waiting at heaven's door.

Everything that I went through was meant to strengthen my soul.
I had to fall all the way down in my present
state, in order that I grow.

But God has a plan for us, whatever we're going through.
No matter what it takes, He'll find a way to help us accept the truth.

They said that I was heartless, and I didn't care who I was hurting,
little did they know that during this time, it was
God who was doing all the working.

It's amazing how we've created nothing, but
we think we have the right to judge.
We even think that we can dictate the
actions of the one who created us.

Understand that when you're someone's number
one focus; it's because they don't have a life,
and they may be doing just what they're accusing
you of in the cover of the night.

You see a true Christian doesn't judge, they pray.
That God removes all of our sins away.

Remember none of us is promised to be here tomorrow.
So be very careful of whom you choose to follow.

Busy bodies and troublemakers have no place in the church.
So stop all this unnecessary gossiping and
let your actions do your work.

From the pews to the pulpits we should all be on one accord.
With nothing but love for one another as
we worship and praise the Lord.

Silent Prayers

My divine creator, Father of Abraham, Isaac, and Jacob.
To whom I owe my very soul for a love that is unconditional.

To my God who created the universe and
everything that lies within,
I come to you from the depths of my heart,
to ask for forgiveness of my sins.

Through you Father God, anything is possible, if it is your will.
I come to you in prayer seeking salvation,
which I know you are ready to give.

As I come to you with open arms, I ask that you relieve
me of all my transgression. And through the strength
that you gave me, my life can only have progression.

And I pray that you bring me and my
enemies together in agreement.
To release all hostile feelings with nothing but contentment.

Most of all, my Father, give me the will to want to change.
Because in my heart I know death is near,
if my life remains the same.

I'm tired of running from you, Lord; I'm
through with my foolish pride.
I know I can no longer make it without you by my side.

I pray for those who are still suffering, who
are in worse shape than me.
Give them the strength to overcome all obstacles,
and the power to make the devil flee.

So from my heart I thank you, Lord, for the
many blessings you have given.
Only through your grace and will that I am still living.

And I thank you, Father, for putting the right people in my life.
Who are pure in heart and dedicated to Christ.

Most of all, I thank you Lord because you
are the one who truly cared.
And Whenever I am down and out and need help. you
always answer my silent prayers.

If It Ain't Christ, then It Ain't Right

If you're not talking about Jesus then I don't want to hear it.
You have no idea what I am dealing with.

I don't have time for nothing else in my life.
All I want to do is live right for Christ.

I've fallen for Satan's tricks many times before.
Enough to know not to let just anyone thru my door.

So if you don't have anything positive to say,
I'm begging you to please, please stay out of my way.

There's nothing more important to me than staying in God's grace.
Even if that means putting you in your place.

Out with the old and in with the new.
Once Christ entered my life, He changed the rules.

I no longer want to follow the examples of this world.
And no longer am I looking at my brothers
with a heart that doesn't care.

I thank God for forgiving me for my sins.
Because I know that I'm being judged for what
I'm doing now, instead of what I did then.

How can you be disobedient to God and keep a sane state of mind.
Do you know that you're running out of time?

And why are you wasting your breath talking
about your sisters and brothers?
Do you not know that God puts us here to care for each another?

I'm pleading with you from my heart before it is too late.
To call on Jesus and let Him put a smile on your face.

He's standing at the door just waiting for you to open.
But before He can enter, you must first rid yourself of sin.

You have to believe in your heart that His word is true.
And that there's no limit to what He's able to do.

So the next time you want to go to someone with that nonsense.
Think twice about what you are doing and then repent.

When you walk with Jesus, you're representing love.
Cast down straight from the heavens above.

And everyone around you will see your light.
Which may influence them to walk with Christ.

What Happened to Those Days

What happened to those days when family
looked out for one another.
When kids grew up showing respect for their mothers and fathers.

What happened to those days when people
got married before they had children.
Which is how God intended it to be when
He created us in the beginning.

What happened to those days when praying
was allowed in those schools.
And kids had no problem with following the rules.

What happened to those days when you
could sleep with the doors unlocked.
And you didn't need security to watch your house around the clock.

What happened to the support that was coming from the church.
When there was help for families that were in need and out of work.

What happened to the marriages that lasted until death do us part.
When couples were getting married because
they both had love inside their hearts.

What happened to those days when a man's words were His bond.
Now it seems that nothing but lies come out of a man's tongue.

What happened to the unity while serving a cause.
And who made the decision to put righteousness on pause.

What happened to not being afraid of dying for what is right.
In those days people were ashamed of their dirty
laundry being brought into the light.

Look at men mating with men and women mating with women.
Convinced by Satan's lies that they are not even sinning.

What happened to those days when man was the breadwinner.
He at least took care of his family, even if he was the biggest sinner.

Maybe if we could playback the tape,
we'd see where we went wrong.
And start worshipping our true Savior who sits up on the throne.

He died on the cross to spare us from suffering,
And there's no better way to demonstrate
the true meaning of loving.

Teardrops

In the aftermath of birth, tears are shed.
The rivers even flow as two are wed.

As wet drops form from the cheeks.
When one can attain just what one seeks.

Enlightenment, happiness, and joy can start the water to run.
Which is sometimes sheltered so as not to be seen by some.

Even in anger, the eyes can swell up into a mist.
To foretell the stories of all the problems that exist.

But the question is whether or not it's healthy to hold it all within.
Which can only take us back to the places we've already been.

Shedding a tear out of joy or grief is a way to express.
That you're handling your emotions no more or no less.

But at that moment you're being truly who you are.
A victim of circumstances, a casualty of war.

You're No Better than Me

What makes you think you're better than me; is it the clothes I wear.

You know nothing at all about me why should you even care.

I may not have the best job, and sometimes I even go without food.
My life goes up and down like a season, depending on the mood.

But I do believe that if you look hard enough,
you'll find good in everyone you meet.
How do you know if that person does or doesn't
have goals they're trying to reach.

I believe that when you find fault in another;
it's a reflection of how you see yourself.
And you have issues going on in your life that are quietly kept.

The only way to make you feel good for
the wrong that you are doing.
Is to target someone's life with the intent to ruin.

Why can't you see that all your problems start from within.
What matters is where you are going, not where you have been.

So take a look in the mirror, and see what comes to your mind.
Stop casting judgment on others and leave the past behind.

If you have something against another, let
it be for something they've done.
And stop thinking that you are better we're
all God's daughters and sons.

We Can Change Hate to Love

It use to hurt when I walked the streets and someone called me a crack head.
I knew what they were saying was right, I just didn't like hearing it said.

In reality they were strangers who didn't know anything about me.
All they knew is what they saw when I walked up and down the streets.

Even my family didn't know what I was going thru.
Because when I was around them, I always hid the truth.

The worst thing that happened was when I began to hate myself.
Once I started doing that, it was easy to hate someone else.

The Bible tells us that we should love each other and always spread joy.
Not to walk around hating others with the intent to kill and destroy.

So I had to stop hating people because of the things they say.
And learn how to work on myself, each and every day.

I had to have forgiveness in my heart for whoever did me wrong.
It wasn't until I started forgiving that I became strong.

It's easy to hate someone whom we don't understand.
Especially when we're ignoring God's words, but listening to man.

Hate is nothing but the devil's work.
It'll make you lie, steal, or shoot up a church.

Hate can poison our spirits; it can condemn our souls.
It'll only lead us down the wrong road.
God loves us all no matter who we are.
We have to learn to love Him back, if we want a changed heart.

If we dissect our hearts and find the slightest trace of hate.
Then we should ask God for deliverance before it's too late.

And if we begin to have a relationship with Father God above.
I guarantee you that we can change hate to love.

Spirituality

I was seeking religion, which is why I went down that road.
Filled with insecurities and doubts about things I didn't know.

I thought drinking and doing drugs was
a social event to be proud of.
Too busy searching for religion, having
no idea what spirituality was.

Instead my life was total chaos, everything I did fail.
Not knowing that by seeking religion, I was headed toward hell.

Then one day God opened my eyes, and I
found myself with a changed heart.
Instead of being the problem in this world,
it's time for me to do my part.

And stop seeking religion which can only
get me sent to the fiery pits.
Have a close relationship with God and focus on my spirit.

In My Dream

When we first meet a stranger, it doesn't matter if their tall, short,
   fat, or thin
But what matters most is the color of their skin.

Why can't we all just get along.
And have peace and harmony in our lives until Jesus calls us home.

In my dream, one day all races will walk hand and hand.
Not judging nor caring about the color of the man.

In my dream we'll find that we have enough love in our hearts.
To stand up against ignorance and bigotry that's trying to tear this
   world apart.

In my dream, one day our children will realize.
That hard work and dedication is the key to thrive.

In my dream God's message will touch each and every one of our
   ears.
And no man will ever again have to face racial fears.

In my dream, one day all of our pains and sorrows will disappear.
And everyone's tongue will confess that Jesus Christ is the cure.

In my dream, the rich and the fortunate will stop bragging.
And our kids will see just how disrespectful they look walking these
   streets sagging.

In my dream, mothers and daughters, sons and fathers will no longer
   fight.
Because we'll be turning all of our battles over to Christ.

In my dream we'll put all of our pride and jealousy to the side.
And embrace each other with nothing but love, as we prepare for our
heavenly ride.

In my dream, there will come a time.
When there will be no more senseless black-on-black crime.

And if we stop judging others for their skin color and start letting
God have His way.
Maybe then the only color we'll see, will be shades of gray.

How We Finish the Race
not How We Start

When I think about where God brought me from, it brings tears to
    my eyes.
I think God for saving me, without Him I would've died.

I was a liar, a cheater, a dope fiend, and a drunk.
I couldn't believe just how low my life had sunk.
I was in and out of jail so much that it became my second home.
And my stinking thinking led me to believe that I did no wrong.
I partied all night at clubs, fornicating with different women.
Every part of my life was somehow connected with sinning.

Yet through all my evil actions, Jesus never left my side.
And I would have turned to Him a long time ago, if it wasn't for my
    foolish pride.

I had to realize for myself that there was nothing for me out on those
    streets.
And if I didn't turn my life over to Jesus, my life would be brief.

The devil had my mind so twisted and confused; I couldn't see the
    truth.
I was so self-absorbed, full of pity because of what I was putting
    myself thru

I didn't feel that I had a problem; I didn't feel that I had to change.
Especially when I had everyone else in my life to blame.

People in my life were being hurt, and I didn't even care.
Each day I had was a living nightmare.

I thank God for giving me a family that knew how to pray.
It was because of their prayers that I believe that I was saved.

If you haven't been filled by the Holy Spirit, you have no idea what you are missing
It'll leave you with a faith so strong, that you'll never again be guessing.

You'll know without a doubt, that only through Jesus can you receive eternal salvation.

This should be enough to feel our hearts up, with appreciation.
I have a ways to go in my walk with Christ, but I thank God that I'm not where I was
And I know that the only reason that I'm still alive is because of the goodness of His love

So if you are a liar, cheater, dope fiend, or a drunk, just accept Jesus in your heart
There's nothing that He can't do; remember it's how we finish, not where we start
You see, the devil wants our heartaches to last.
That's why He's trying so hard to keep us in the past.

Because He knows that it can only make us strong.
When we forgive ourselves and others and just move on.

He's going to try it again, again, and again.
To keep us down and out because of our past sins.
So if someone's still holding a grudge because of what we've done years ago.
Just put that person in God's hand, and let Him carry your load.
When God sees us, He's looking straight at our heart.
He cares about how we finish the race, it's not about where we start.

Worship Only

When we come to the house of God, we should have worship on our
    mind.
And leave all of our negative thoughts and feelings behind.

All the pagers and cell phones should be turned off.
If we don't pay attention to God's message, our souls could be lost.

God deserves one hundred percent of our attention.
After all God's messengers are a part of His extension.

If we're going to church for any reason other than worshipping the
    Lord.
Then maybe we don't have Jesus in our hearts.

The attitude that we bring to church has to be humble and tamed.
We have to take God seriously; salvation is no game.

We can't use the church for our own selfish gains.
We were created to serve God and give praises to His name.

God knows what's in our hearts; He knows what we're about.
That's why He's always one step ahead of us, to show us a way out.

But with the wrong attitude, we'll think that we're always right.
And that's how we distance ourselves from Jesus Christ.

Seek first the kingdom of heaven, and our souls will be saved.
All because of Jesus's love and the life that He gave.

If you accepted Jesus as your savior, you'll truly believe.
That the house of the Lord is for worship only.

Patience

Why am I always constantly putting myself through heartaches, pains, and sorrow.

Wanting love so badly in my life, that I'm willing to sacrifice my tomorrows.
What is this about this thing called love that always leaves me holding the bag
And going thru life being afraid of it continually keeps me sad.

I've come to a point in my life, where I've started looking within.
In hopes of finding out just when this problem began.

I know that I have a lot to give; all I want is the same in return.
Every time I think that I've found it, my heart just ends up burned.

So am I destined to live my life alone.
With so much love I have to offer, I know that this is wrong.

Or am I expecting too much out of another.
All I want is just for us to love each other.

I know deep in my heart that if I continue to pray.
God will one day send me my true mate.

All I have to do is don't give up and keep love in my heart.
So that when she does come along, I am ready to do my part.

Maybe all my past relationships were all meant to fail.
So that when she does come into my life from my heart I'll be able to tell.

That true love does exist if you let God choose your mate.
And that can only happen if we have patience and faith.

Made in His Likeness

If I wrote a book about me what would it be.
Comedy central, Syfy, Western, or would it be Lifetime TV.

If I wrote a book about me, would I tell a lie.
Because the truth always seems to make me cry.

Or would I find the courage to tell the truth.
So that everyone will know how it feels to walk in my shoes.

They say that you can't judge a book by its cover.
Yet whole nations of races are being condemned just because of their
color.

It's amazing how people can take how you look and turn it against
you.
Your hands, your toes, your lips, your nose are some of the things
that they use

They will even talk about how you eat or how you sleep.
Or maybe they don't like the shape of your teeth or the size of your
feet.

God said that He made us all in His own likeness.
Which means no matter how we look, we are all full of His greatness.

What's more important is how we feel about ourselves.
If God loves us just as we are, then we shouldn't worry about how no
one else felt.

After all why would angels go against God because of us.
If we didn't possess God's great love.

If I wrote a book about me would I use my real name.
Knowing that there's a great price to pay for earthly fame.

If I wrote a book about me would I tell lies to another.
Just because I want to come out looking better than my brother.

It's so easy for others to get into our minds and beat us down.
All they want to do is turn our smiles into a frown.

Constantly reminding us of the disappointments we've had in our
    lives.
Hating when we laugh and loving when we cry.

Just remember how they beat Jesus down.
They put thorns on His head and called it a crown.

And even though He knew exactly what was in store.
He died so that we didn't have to worry about our souls anymore.

Negativity is an addiction that doesn't want to let go.
And if we let it hang around long enough it could poison our soul.

Remember we are all made in His likeness.
Which means no matter how we look, we are full of His greatness.

I've Finally Come Home

I spent most of my time doing things that I knew were wrong.
Which is why I've always felt alone.

But when I turned my life over to Jesus, I became strong
Because only thru Him was I able to come back home.

I use to think that living for this world was what I wanted to do.
Regardless of the pain and suffering that it put me thru.

Nothing that I did ever turned out right.
But I kept doing things my way, neglecting Christ.

Many nights when I was out there strung out on drugs.
I forgot about decency; I forgot about love.

The devil had me convinced that I didn't need anyone in my life.
He had me thinking that I could survive without Christ.

He didn't want me to know that Jesus's love would always stand
    strong.
That's why he always tried to keep me alone.

When God delivered me from myself, and unlocked the doors of my
    prison cell.
Not only did He free me from confinement, but He also rescued me
    from hell.

He revealed to me the limitation that had my life on hold.
And gave me the courage and the strength to be bold.

He also showed me the responsibilities that I have for my sisters and
    brothers.
And how wrong it was to cast judgement on others.

When it comes to following His rules, there's no compromise.
If I want the kingdom of heaven to be my ultimate prize.

Now I know without a doubt that I will never be alone.
Because I've invited Jesus in my life, and I've finally come home.

What in Hell Do You Want

It's time we make a decision about whom we're going to serve.
Everything about heaven and hell is written in His word.

So what in hell do we want that's better than salvation.
God has shown us that we needed Him
since the beginning of creation.

But no matter what He does for us, we've always turned our backs.
Which is a perfect illustration of the faith that we lack.

If our name is not written in the Book of Life,
we're doomed to eternal suffering.
And all our begging and pleading will only lead to nothing.

At that moment, if judged wrong, we'll
be thrown into the fiery pits.
Having been judged only by our actions on our sinful list.

Just repent for our sins, and believe that Jesus died on the cross.
There's nothing that we shouldn't do to
prevent our soul from being lost.

There's nothing in comparison to the
blessings He has in store for us.
So tell me what in hell do you want that's better than God's love.

All Satan wants to do is use us and then kick us to the side.
And if we think we can live a life of sin and
still go to heaven, we're living a lie.

My God

I'm not bragging, but let me tell you what my God has done for me.
I was born in this world full of poverty.
Yet God has always provided for me.

When I was bedridden from a stroke, He set me back up on two feet.
He protected me from racial injustices and gave me my dignity.
I'm not bragging but let me tell you what my GOD has done for me.
When they said that I was illiterate, He taught me how to read.
And when I was in a losing situation, He gave me the victory.
When I thought that I was all that, He taught me humility.

He stopped me from being a follower and showed me how to lead.

I'm not bragging, but let me tell you what GOD has done for me
He gave me a loving and caring wife.
He brought my children back into my life.

When I was stressed out and on the verge of mental breakdown, He
gave me inner peace.
By first teaching me how to get down and worship Him on bended
knees.
I'm not bragging, but let me tell you what my GOD has done for me.

When I was dying from an overdose He breathed life back inside of
me.
He delivered me from a fifteen-year crack addiction and from being
a menace on the streets.

He brought me from a prison cell and showed me how to appreciate
being free.
And put a roof over my head to keep me from sleeping on the streets.

I'm not bragging but let me tell you what GOD has done for me.
He protected me during war on a Navy ship in the middle of the sea.
And He always strengthens my soul, when my faith felt weak.
When I am hungry He always provides me with something to eat.
And when I saw no way out, He always paid those bills for me.

I'm not bragging, but let me tell you what my GOD has done for me.
He registered me a place in heaven throughout eternity.
And no matter what my faults are He still loves me.
He's the Alpha and the Omega, the divine Trilogy.
I'm not bragging, but let me tell you what God has done for me.

As Long as I Am Still Alive

Even though I am homeless, I thank God that I am still alive.
Because I know that without problems there would be no need to
strive.

I may not have fine clothes, and I may not be in good health.
But as long as I have Jesus, I have no need for nothing else.

I've come head to head with death many, many times before.
I know that one day my suffering will end, and I'll walk thru heaven's
door.

And no matter what I'M going through I know that God is by my
side.
And it's a blessing and a privilege just to be alive.

I feel it deep inside my heart, that Jesus is coming soon.
Which is why I'm constantly praying, with my heart and soul in
tune.

What good is it to gain this world, but lose my only soul.
When I can accept Jesus in my life, and make heaven my number
one goal.

When I was strung out on drugs, I felt like I was running out of time.
And then Jesus reached out and held me in His arms and gave me a
peace of mind.

When I was out there in the streets, God never left my side.
And it's only because of His love for me, that I am still alive.

It was God who put the right people in my path to help me go straight.
They helped turn my life into a miracle, when I thought it was too
late.

So it doesn't matter if I'm troubled, homeless, dirty, or broke.
Because as long as I have Jesus, He'll always give me hope.

When I thought I had everything, I really had nothing.
And now when I Haven't a penny to my name, I truly have something.

So no matter what tomorrow may bring, there's one thing that I can't
    deny.
I can change, I can make a difference, as long as I'm still alive.

Let Jesus Choose Your Friends

As a young child, I grew up with drug dealers across the streets, pros-
titutes next door, and alcoholics in my own home.
This is what I experienced in my life from a child until I was grown.
There were parties that ended up in fights; they all ended up the
same.
People showing out in front of others, just to make themselves a
name.

And the movies that I watched did nothing but add to my struggles.
They had me believing that I had to prove myself to others.

In the neighborhood that I grew up, you had to know how to fight.
Hanging out on the corners during the daytime and breaking into
cars at night.

But the older that I got, I came to realize that my life was all about
excuses.
I had to understand that I was responsible for my own actions regardless of my abuses.
When I got in trouble where were those so-called friends when I need
them the most.

As soon as I got locked away I immediately became a ghost.
I asked myself, "Am I a product of my environment or am I just weak
within?"
There has to be a reason why I'm making the same mistakes, again
and again.

Is it just plain ignorance or is it a lack of control.
I know that I wasn't born, with a twisted and confused soul.

There are rewards for doing what's right, so is it patience that I lack.
Why is it so hard for me to just give in and just accept the facts.
That without Jesus in my life, I will continue to go on that merry go
 around.
Waking up each morning with my soul in the lost and found.
The struggles in my life will never come to an end.
If I don't sit back, and just let Jesus choose my friends.

There is no excuse for being raised in your environment.
It's all about where we are going, not where we have been.

I know old habits are hard to break and change can seem hard at times.
But if we put our trust in Jesus, He'll give us a peace of mind.

It comes a time when the generation curse must be broken.
But first we have to keep our hearts and our minds open.

Our lives would be nothing but a waste.
If we walked around full of hate.

And just because our mothers and fathers choose to live life the
 wrong way.
Doesn't mean that we are destined to make the same mistakes.

The main ingredient to being a successful woman or man.
Is to just sit back and let Jesus choose your friends.

My Testimony

There are many addictions in our life that we have to face each day.
And some of these addictions may have the power to lead us astray.

Not all addictions will lead us toward alcohol or drugs.
Some of us are addicted to lust, power, control, and love.

We fool ourselves into believing that we have to have someone in our life.
Not even caring about whether or not the relationship is wrong or right.

We love bossing others around and telling them what they need to do.
But when it comes to admitting our own faults we can't accept the truth.

And then there are those of us who have a need to shop until we drop.
Because when it comes to spending money, we don't know how to stop.
Just remember that an addiction is nothing more than just a habit.
Constantly trying to convince our minds that we just got to have it.

To tell the truth, we all have something that we need to be delivered from.
It could even be caffeine, gambling, jealousy, or bad language that comes from our tongue.

So let me tell you how I beat my addiction of smoking crack cocaine.
My addiction took away my self-esteem and left me feeling insane.

I was in and out of jail so much that I called it my second home.
Crack isolated me from my loved ones because it wanted me all alone.

Through heartaches and pains, I lost those that were close to me,
    whose faces I'll never see again.
My addiction told me that I didn't need anyone; it told me that it
    was my best friend.

Crack called me when I was woke, and it haunted me in my dreams.
My whole body felt like I was being pulled by a tractor beam.

But in reality, I was like a scared little child left alone in the dark.
Mad at the world and blaming others because I couldn't live up to
    my part.

And each time I hit that pipe, it took a part of my life away.
Crack just wouldn't leave me alone; it told me that it was here to stay.

It's amazing how something can start off good, but end up so bad.
It can take control over everything else in our life and leave us feeling
    sad.

It'll have us running and hiding, too afraid to face anyone else.
Destroying everything and everybody in our lives, until there is noth-
    ing left.

I was too foolish to admit my wrong, and too prideful to ask for help.
My mind, body, and soul felt lost; I was one step away from death.

That was when I saw my life flash right before my eyes.
And I couldn't take it any longer, so I just broke down and cried.

I cried out, "Please, Lord, help me. I need you; take control over my
    life.
Come into my heart, my Lord, and help me to live right."

Even though I didn't hear a response, I had enough faith to believe.
That if I turned my life over to Jesus, He would be all that I need.

My testimony is to tell everyone exactly what I witnessed.
That if Jesus hadn't come into my life, I would still be crack-addicted.

And if He can deliver me from a fifteen-year addiction to crack
cocaine.
I believe with all my heart that He can treat your addictions just the
same.
So no matter what we are going through, God can make a change in
our life.
We can't do it alone; we have to have Jesus Christ. And remember
that an addiction is nothing more than a habit. With the enemy
trying to convince us each day that we just got to have it.

Now I thank God every day down on my knees.
Because of His love for me, I have this testimony.
There was a time when I fell to my knees and closed my eyes and
said, "God kill me or save me, I can't take no more."
And when I opened my eyes Jesus was standing at my door.

He Gave Me Back My Life

If we create our demons, then only we can chase them away.
But first we have to stop jumping to conclusions about what people
   think or say.

None of us have the power to read what is going on in another person's mind.
So when we get upset about something that we don't understand
   we're actually walking blind.

That's why every time we get a bad thought about someone, we have
   to rebuke the devil in the name of Jesus Christ.
That's the only way that we're going to feel peace and have prosperity
   in our life.
When some of us get a little money we get caught up on our high
   horses.

We need to always know that pride and greed are destructive forces.
It can destroy a good family in a single day.
It could run a very good friend clean away.

Pride can cause us to feel like we have nothing to repent.
While greed can cause us to miss out on a blessing that's heavenly
   sent.

And if we're not careful we'll start lying to ourself.
Just because we have nice clothes and a nice car doesn't make us better than anyone else.

God don't judge us by the clothes on our back or the money in our
   bank account.
He judges us by our actions and by the words that come out of our
   mouths.

How can we love God if we're more in love with material things in
    our life.
God loves us so much that He gave us His son, Jesus Christ.

So if I wrote a book, I couldn't write about me.
Without first mentioning how Jesus came into my life and set me
    free.

How He took the limitations out of my life.
When He taught me how to live right.

Now I have no doubts in my mind that I will succeed.
I will walk thru heaven with Jesus one day, just because I believe.
When I'm feeling low and sad and all alone.
He knows exactly what to do to make me strong.

Oh, what a wonderful feeling to have Jesus in my life.
Watching over me, protecting me, morning, noon, and night.

While I was living in a world of sin and not caring about where it
    would take me.
I almost sent my soul to hell throughout eternity.

You see the devil did everything in his power to take me out.
And he almost did, until the name of Jesus came out of my mouth.

It was never meant for any one of us to spend eternal life in hell.
But the devil has been trying to take us all there, since the moment
    that he fell

And I am here to tell you that the devil is a lie.
As long as we have Jesus in our life, we don't ever have to give up or
    cry.
Now every time that I feel like I am in need.
I talk to my savior, Jesus, and He's always there for me.

And I know that the only way that the joy I feel in my heart will last.
Is if I keep on moving forward and don't let anyone take me back to
    my past.

It's amazing how God can forgive us for things that we've done, but
    we can't forgive ourselves.
And some of us are even too ashamed to ask God for His help.

I'm not bragging, but let me tell you what God has done for me.
He brought me out of a losing situation and gave me the victory.
I was homeless, depressed, and on drugs, and I was financially broke.
And because my lifestyle was so reckless, I even had a stroke.
The devil wanted me to go through all of this alone.
He was trying hard to make hell my home.
But one rainy day I got down on my knees and called out Jesus's
    name.
And whether I knew it or not, my life was about to change.
So now every time the enemy tries to take me back to my past.
I call on Jesus and I just sit back and laugh.
And I know that if Jesus could save me from all the bad things that
    I went through.
If you really want a change in your life, just talk to Him, He'll do the
    same for you.
When I decided to throw my hands up in the air and accepted Jesus
    Christ.
From that moment on He gave me back my life.

We Are All Special

Have you ever wanted to be someone else.
Because you were too ashamed of being yourself.

And you treated life like it was some kind of game.
Stepping in and out of your very own lane.

Taking the gift that God gave you and put it to the side.
Just because someone else offered you a different kind of ride.

The last thing the devil wants us to do is to follow our dreams.
He wants us to forget about what dedication means.

That way he can keep our minds twisted and confused.
Not even realizing that we are being used.

That's why whatever we are doing we have to involve God in our life.
He's the only one who's going to lead us down a path that's right.

We have to be very careful about whom we choose to call our friends.
We can't just go around trusting any and all men.

We can't keep involving the wrong people in our life.
If we plan on serving Jesus and living right.

We have to let God make all of our decisions.
But first we have to seek the Holy Ghost and stop chasing religions.

And believe that Jesus died and resurrected in three days.
And that our price for salvation has already been paid.

That's why Christians should associate themselves with those who
     share their same beliefs.
Someone who believes in love, justice, honor, and peace.

Someone who's always willing to do their part.
And someone who's not afraid to have a forgiving heart.

We want God to forgive us for the sins that we commit.
But when someone does us wrong we can't seem to forget.

We want God to come to our aid when we're down and in need.
But when it comes to helping others we're not willing to plant good
    seeds.

We want God to give us a healthy and fruitful life.
But we have a problem with treating others right.

We want someone in our life to grow old with and to love.
Not caring if it comes from hell or heaven above.

God wants all of us to have a happy and fruitful life.
That's one of the reasons He sent us Jesus Christ.

The Bible tells us that a joyful heart is good medicine for the body.
But if we want joy and peace in our lives, we have to live godly.

God knows our weaknesses; He knows our mistakes.
That's why He's always making a way for us, no matter what it takes.

We should never feel like a fool, just because we have a good heart.
Helping others, wanting nothing in return always pleases our Lord.

It's time for us to stop worrying and listening to someone else.
If God haven't given up on us, then we shouldn't give up on ourselves.

Each one of us was born with our very own special gift.
Something that we are especially good at, that gives our lives a lift.

All we have to do is ask God, He'll open up our eyes.

That way we'll see who our true friends are, we'll see right through the devil's disguise.

God is waiting to bless us, but we have to take the first step.

That's why He made each one of us special, so that we don't have to want to be no one else.

Benefits of Working for Jesus Christ

Some people don't understand us when we hold our heads up when
they should be down.
It confuses them to see us with a smile when we should have a frown.

All they have to do is ask, we'll tell them the reason.
we have enough faith that God will give us a better season.

Everything doesn't always go the way that we want them to.
And that's why we have to be obedient and learn how to follow God's
rules.

Whatever we're going thru in life we can't make it alone.
We all need God in our life, if we want to be strong.

But just because we follow God's rules, doesn't mean we won't have
our ups and downs.
There'll always be some issues, as long as we stand on earthly ground.

That's why we always have to remember where God brought us from.
If it wasn't for Him, we never would've gotten through our storms.

How He picked up our children and brought them home from the
streets.
And how He gave us strength when our bodies were feeling weak.

We say that God is good all the time, but do we believe it.
Then why do we worry and cry when we're going thru it.

Why do we turn our backs on others when they are in need.
When we know that that's how we plant good seeds.

We need to encourage one another and lift each other up.
If we want God to spiritually fill our cup.

And stop being afraid to express how we love.
Especially when we're being showered with it from our heavenly
Father above.

We all know that God wants us to first love Him.
But when was the last time we told someone else that we also love
them.

When was the last time that we put someone else's life before our
own.
and opened up our arms, and invited them into our home.

Anybody can look good and holy when they're sitting in church.
But how many of us want to go out in this world and do God's work.

It's up to all of us to get out there and do our part.
Remember God doesn't care about how we look on the outside; He's
looking at our heart.

If we really want to call ourselves God's disciples.
Then we have to start telling people about what's written in the Bible.

I'm not talking about standing in a corner and calling everybody a
sinner.
I'm talking about letting folks know that if they follow God, they will
be a winner.

Let them know about the goodness in heaven and what's in store.
If they invite Jesus into their lives, they won't have to worry no more.

Give them your testimony of how God came in, and changed your
life.
Just because you made up your mind and accepted Jesus Christ.

They should know that if they want to work for God.
That loving and caring for others is a part of their job.

And even though they'll never get a day off.
There are great benefits when you work for the number one boss.

When it comes to representing God, there's no hard work.
It's just as easy as sitting in the church.
Our biggest problem is when we go out in this world, we act like
we're too ashamed.
To talk about the goodness of God, or even mentioned Jesus's name.
We're too afraid someone might not like us anymore.
Instead of letting them know that it's alright to invite Jesus through
their door.
We should be able to live by example, not demand.
That's why we have faith that our children will make the right stance.
God never forces Himself on us, so why do we try to force Him on
others.
We need to stop judging other people because of their struggles.
So instead of sitting around trying to figure out the reason
We need to let them know that we all have our bad season.
And whatever they're going through will pass; it will go away.
As long as they have faith in God, they'll see a better day.
Let them know that all they have to do is ask, and they too can have
a changed life.
But it's up to them if they choose to work for Jesus Christ.

The Cross

No matter what we have are what we've done, we would all still be lost.
If Jesus Christ hadn't died for us that day on that cross.
If there wasn't a resurrection then living would be in vain.
And not one of us would ever have anything to gain.
People are always trying to put themselves into God's shoes.
Always asking themselves or others, what would Jesus do.
Would He want us to forgive or would He want us to get even.
But what it really boils down to is what do we believe in.
"Do we believe that Jesus was the son of God?" His own disciples
    asked that question.
Or was He just another prophet here to teach us a lesson.
Even while He was being crucified He asked God to forgive His enemies.
And He did that even though they didn't believe.

So do we believe that Jesus died on the cross, and in three days came
    back alive
Or are we going to let the enemy convince us that the resurrection
    was a lie.

There was a thief who spoke up for Jesus as they were both dying on
    the cross.
And Jesus told him that because of his faith, that his soul wouldn't
    be lost.

Jesus's words were, "Forgive them, Father, for they know not what
    they do."
And even while He was dying on that cross, He was praying for for-
    giveness for me and you.

Peter once asked Jesus, how many times should He forgive His
    brother.
He asked, "Was seven times enough?" but Jesus told Him that His
    forgiveness should go much farther.

Jesus was telling us that there should be no limit to how many times
that we should forgive one another.
After all, didn't He tell us that we should always love each other.

Not Buddha nor Muhammed nor Confucius died and came back to
life in three days.
And that alone should tell us that only Jesus has the power to save.

After His death, his disciples saw Him from afar and couldn't believe
their sight.
Sitting at the edge of the waters preparing a meal was the one and
only Jesus Christ.

We think that God has abandoned us, yet He never leaves our side.
It's us who separate ourselves from Him, when we get caught up in
greed and pride.

It's always a joy to God when He showers us with blessings
He wants us to understand that everything we go thru has some kind
of lesson.

Jeremiah 29:13 says, "You shall seek me, and find me, when ye shall
search for me with all your heart."
God is just waiting on us to come to Him; He's always doing His
part.

Jesus said that the second greatest commandment is to love others as
we love ourselves.
He didn't say nothing about anyone being better than anyone else.

Remember if we're doing something for someone, and it has to be
mentioned.
Then we may just be doing the right thing but with the wrong
intention.

Some of us brag every time we help out someone else.

Not even giving God the glory, even though it's Him who's supplying the wealth.

Jesus said, "I am the bread of life; He that cometh to me shall never hunger and He that believeth on me shall never thirst."

He's telling us that we can have whatever we ask for, as long as we put Him first.

First John 5:14 says, "And this is the confidence that we have in Him, that if we ask anything according to His will, He heareth us."

That alone should tell us about the goodness of God's love.

God doesn't want anyone of us to be unhappy; He doesn't want none of us to be lost.

He loves us so much that He sent His son Jesus to die for us on the cross.

Jesus Is the Answer

The mind controls the body, but the spirit controls the mind.

While the mind keeps us in the past, the spirit teaches us to leave the past behind.

The mind tells us to hate anybody that does us wrong.
But the spirit teaches us that forgiveness is what makes us strong.

And if God wants His message to be heard.
He doesn't need to use a thousand words.

God can use anyone that He wants to spread His good news.
That's why we have to give everybody a chance to be used.

Some of us talk so much because we are in love with our own voices.
Blocking out God's other messengers and ignoring God's other choices.

God can use anyone that He wants, after all He is the creator.
And He's not making His choice just because someone is a better orator.

Moses had a stuttering problem, but God sent him to talk to a king.
And David was able to calm down bad spirits because God blessed him to sing.

So never let anyone tell you that your life has no meaning.
God has a purpose for all of us; we just got to stop dreaming.

When we think that our purpose is greater than others.
Then we are overlooking another person's struggles.

And just because someone else's purpose doesn't match up with ours.
That doesn't mean that God gave us more power.

When Judas first started following Jesus, He didn't know he'd betray
    Him in the end.
And even though he walked and talked with Jesus, he still couldn't
    escape that sin.

At twelve Jesus was teaching men five times His age.
Which shows that God can use anyone to teach others how to behave.

We think that we know things that we can't understand.
I don't care how smart we get; we'll never know God's master plan.

I don't care if we're broke, homeless, lonely, sick, or dying from cancer.
If we want everlasting life, we have to believe that Jesus is the answer.

Jesus is the answer to keeping our family together.
Jesus is the answer to making our lives better.

Jesus is the answer to that addiction that's keeping us on the run.
Jesus is the answer to that sickness that we are dying from.

Jesus is the answer to that lonely and broken heart.
Jesus is the answer to that change in our lives when we're searching
    for a new start.

Jesus is the answer when we are feeling down and weak.
Jesus is the answer when we've lost our children to the streets.

Jesus is the answer when we are dealing with bad health.
Jesus is the answer when we've lost a loved one due to death.

Jesus is the answer when we are hungry for bread.
Jesus is the answer when we need a roof over our head.

Jesus is the answer when our marriage is falling apart.
Jesus is the answer when we need love back in our heart.

Jesus is the answer when we're broken and out of work.
Jesus is the answer when we are crying because we are emotionally hurt.
And if we believe that Jesus is the answer, we need to get out of His
    way.
And give Him praise each and every day.

God Is Good All the Time

When my body aches with pain and my heart is filled with sorrow.
I still have faith that God will comfort my tomorrows.

I know that God will never turn His back on me no matter what I'm
going thru.
Because I believe that when He says that He loves me, He's telling
the truth.

I also know that my pains could never match those that Jesus endured.
Which shows me that the love that He has for me is strong and pure.

Jesus says that if my faith was the size of a mustard seed I could move
mountains.
Which is why I believeth that God can deliver me from any kind of
sin.

God wants me to love a stranger just as much as I love myself.
He wants me to appreciate life and stop being afraid of death.

He wants me to speak respectfully and boldly when I call out His
name.
He wants me to know that there will be a day when I feel no pain.

He also told me that whatever my mind thinks, my body will feel.
If my mind tells me that I'm sick then my body would feel ill.

If my mind tells me that I'm unhappy, then my spirit will feel sad.
If my mind tells me that I'm upset, it'll convince my spirit to feel
mad.

And that's why I have to stay prayed up each and every day.
I know that if I don't fill my heart up with love, it will be consumed
by hate.

And most importantly I have to realize that I'm not perfect; I make
    mistakes.
And the day that I start lying to myself is the day that I become fake.

Sometimes we say things that we really don't mean.
And sometimes things ain't always what it seems.

We don't know what's going on in another person's mind.
And that's why we still have to love them, even when they cross that
    line.

Oh, like you've never woke up and had a bad day.
And you didn't want to hear what anybody had to say.

If we don't like someone, we can find fault in everything they do.
Hatred is one thing that'll make us run away from the truth.

When Peter had supper with the Gentiles, the other disciples couldn't
    understand.
Even though some of them walked with Jesus, they couldn't see God's
    master plan.

The gift of heaven has always been meant for all of us to reach.
And if we want to send out a message of hope, this is what we should
    teach.

Like the woman who knew that she'd be healed just by touching the
    hem of Jesus's clothes.
Because Jesus loved her faith, He made her whole.

Just like when David was running from Saul because he feared his
    death.
Not only did God remove Saul, but he also gave David his wealth.

And look at Moses; he was eighty when God sent him on his mission.
Which goes to show that no one can see God's vision.

Jonah refused to go to Nineveh to deliver God's ultimatum.
Because he knew that God would still forgive them.

Even when Peter cut the soldier's ear off in the garden.
Jesus was able to reach out to him to teach him a lesson.

We are the problem, not God, we cause things to happen to ourselves.
Especially when we put our lives above someone else.
And don't give God the glory when He does amazing things in our
    life.
As soon as our problems are solved, God is no longer in our sight.
We have to make up our minds on whom we are going to serve.
We can't keep kicking God to the curb.
And we should have no problem making up our mind
If we truly believe that God is good all the time.

Big God Equals Little Problem

Husbands and wives, sisters and brothers.
This is not the time for us to be at war with each other.

Whatever issues we are having, we have to believe that God can solve
them.
After all we have to remember that big God equals little problem.

When we are facing adversities and our lives seem lost.
That's when we have to stand strong on our faith and believe that
Jesus has already paid the cost.

No matter what we're going through we'll never understand.
If we can't feel the pain of the nails that were driven thru Jesus's hand.

We have to believe that our lives do matter.
And if we turned our lives over to Jesus, He would make everything
better.

The misery, the pain, and all the strife.
That we have had to go through in our life.

Happened for a reason, it was all according to God's will.
And until we began to believe this, we'll never be able to heal.

It's always a test when trouble and despair choose to come.
And only God can give us peace amidst the storms.

The enemy is doing his best to kill, steal, and destroy.
But Jesus is here to save us and give us joy.

Whatever issues we are having, we have to believe that God can solve
them.
After all we have to remember that big God equals little problem.

We are all somebody, even though some people may not see it that'll
    way.
Which is why we should give thanks to God each and every day.

Doubt is poison to our minds; it will take us to places we can't
    understand.
That's why we have to turn our lives over to God and obey His
    commands.

And stop being afraid of the devil; he has no power.
Why else do you think that he's always sour.

It doesn't matter if we're facing death, ridicule, prejudice, homeless-
    ness, starvation.
God will always come to our aid; we are all His creation.

There are a lot of hypocrites pretending to be holier than thou.
Who better not wait too late to confess their sins with their mouth.

There are those who think that what they do in the dark can't be
    seen.
They need to know that nothing can be hidden from the King.

We can talk about them, call them hypocrites, non-believers, or blind.
Or we can pray for their souls and that God gives them a change of
    mind.

We've got to stop letting others control how we behave.
And ask God to help us with our issues if we are really saved.

God has already equipped us; He has empowered us to not live in
    fear.
And He'll never leave our side; His presence is always near.

God is not trapped inside a closet; God is inside our hearts.
He is the light inside of us outshining the dark.

Ever since this virus has arrived, people are losing their minds.

Jesus is on His way back, and this is just another sign.
So many people on the streets reaching out for someone's help.
But we are afraid of getting close to them because it may mean our
own death.
The enemy may kill our children out there in the streets.
And he can tempt us with the very things in life that make us weak.
But he can't do anything unless God allows Him to do it first.
And that's why we have to turn to Jesus, to quench our thirst.
It doesn't matter what the enemy throws at us; God can always solve
them.
After all, when it comes to God there's no such thing as a big problem.

Faith

Have you ever felt a summer breeze.
And the best part about it is it didn't make you sneeze.

And you felt a peacefulness so deep inside.
You felt so much joy in your heart that it made you cry.

Not tears of sadness but tears of relief.
There's no better feeling than when God takes away our griefs.

We all go through situations in our life that make us worry.
And that doesn't make us no different from others; that's just our
story.

But when we believe in the name of Jesus and the power that it has.
WE have to believe that if we want anything, all we have to do is ask.

After all, didn't God love us enough that He sent us His son, Jesus
Christ.
Who left heaven to die for our sins, so you and I could be introduced
to the light.

The enemy wants to keep our minds occupied with pain and sorrows.
He wants us to give up on our tomorrows.

And cast all of our hopes and our dreams to the side.
He just wants us to stop believing, and give up and die.

He's convinced so many people in this world that God doesn't exist.
He uses lies and deception as his biggest tricks.
The devils going to try and tempt us with something every single day.
He wants our souls to be lost and go astray.

We have to stop letting our problems weigh us down.
And start putting our faith on solid ground.

We have to keep our heads up; everyone makes mistakes.
That's why we have to repent for our sins before it's too late.

God does exist; I don't care about what anybody says.
I feel His presence in my life each and every day.

I used to didn't care about how much I sinned.
And I still would be that way if I hadn't let Jesus in.

When He came into my life I was one step away from death.
He taught me how to love again, by first loving myself.

He delivered me from a life that was completely out of control.
All because He loved me and wanted to save my soul.

Now no matter what bad situation that I find myself in.
I turn my problems over to Jesus when I use to turn to sin.

But I don't think I'm the second coming of Jesus Christ.
There are still some things that I have to be delivered from in my life.

The Bible says that a sin is a sin, not one is greater or lesser than another.
And that's why we have to stay prayed up and be careful how we treat
    each other.

We all have the power to put the devil in his place.
As long as we don't let anyone or anything make us lose our faith.

Just because we're going through something, doesn't mean that God
    has left our side.
Some of us just need a lesson in humility, especially when we're full
    of pride

Everything that happens in our lives is according to God's will.
And that's why it's so important to have faith that He is real.

The Bible tells us that if we have enough faith mountains can be
    moved.
Faith is not only our biggest asset; it's our most valuable tool.

It's the reason why we come to church, the reason why we worship
    and pray.
The reason why we believe we're going to wake up each and every
    day.

It's the reason why we love a God whom we've never seen.
And the reason why we know that our souls have been redeemed.

It's the reason why we pray for those who hate us so.
The reason that we smile when we should be feeling low.

For God so loved the world that He sent His only begotten son.
And when Jesus was nailed to that cross, He knew that He had
    already won.

So the next time the enemy tries to make us fall for his bait.
Rebuke him in the name of Jesus and let him see your faith.

Let Him see the strength that you possess inside.
Let Him know that He no longer has the power to make you cry.

Then watch how fast the enemy run away; He doesn't want no part
    of you.
As long as you let Jesus run your life, the enemy can't tell you what
    to do.

We all have the power to put the enemy in His place.
As long as we don't let anyone or anything make us lose our faith.

If We Don't Use It, We May Lose It

God has given each one of us a special gift.
Something to boost our lives and give our confidence a lift.

Some of us use this gift in our lives each and every day.
While others simply let this gift from God just slip away.

The last thing that Satan wants is for us to reach our highest potential.
He knows that God created us all to be special.

He also knows that God intended for all of us to use it.
And if He can keep us from our destiny He knows that we will lose it.

He wants us to feel as if there's nothing that we can do.
He'll do anything to make us give up on our dreams by running from
the truth.

He's afraid of the power that God has given to each one of us.
He hates it every time God shows us love.

The last thing the devil wants is for us to love ourselves.
Because He knows that that's the first step toward loving someone
else.

Everyone is born for a special reason.
We just have to remain patient and faithful until it's our season.

No matter what our gift is it's up to us to use it.
And that's why a lot of us end up losing it.

First we have to accept our gift and then we have to understand.
That no matter how far our gifts take us in life, God is still in command.

Second we have to believe that our dreams can and will come true.
And third God is going to always be there for us, regardless of what
we're going thru.

He wants to see each and every one of us succeed.
But we can't let our gifts turn us toward greed.

When God gives us a gift, we should never abuse it.
And if we don't use God's gift, we may just lose it.

The Judge Sits on the Throne

The darkness that surrounds me, keeps me engulfed in the night.
But I know that only you, Lord, can shine a light throughout my life.

The guilt that I feel inside my heart every time I commit a sin.
Let's me know that I am in constant battle, spiritually within.

I don't know if it's shame or grief that makes me keep my mouth
closed.
Or maybe there are some things about my past, that only God should
know.

No matter how good their intentions are, men have a tendency to
judge.
And somewhere deep within their hearts, they could even hold a
grudge.

But I am so tired of disappointing you, Lord.
Walking in the footsteps of Jesus sometimes seems too hard.

I want to be perfect, but I feel like I'm setting myself up for failure.
But no matter how many mistakes I make, I'm still gone be a believer.

It's hard for me to look at myself in the mirror when I know I'm
doing wrong.
Especially since I know that God is watching me atop His throne.

But no matter what I do, I know that I have to keep on living.
That's why I'M grateful to serve a God of love, whose heavy on
forgiving.

Maybe I'm just feeling guilty for holding judgement on my brother.
Instead of wondering who's saved and who's not, I should be praying
for their struggles.

How do I know what relationship is going on with God in someone else's life.
It doesn't matter what we see on the outside, the inside could be shining with light.

With all the miracles that Jesus performed, the priests still accused Him of doing wrong.
Which proves that no one is capable of judging others, except for our Father who sits on the throne.

God's Mercy, Grace, and Love

Save me, Lord, from my demons within.
That are doing everything in their power to keep me drowning in sin.

Save me from my addictions, that are keeping me alone.
Give me my freedom Lord, make me strong.

Please put the right people in my life.
Who will lead me toward your everlasting light.

When I am weak, and I am about to fall.
Open up my ears my Lord so that I may hear your call.

I can't make it without you, Lord, that much I know.
Give me wisdom, give me love so that I can spiritually grow.

Regardless of what you do for me, Lord, I know it won't matter.
If I don't have the desire to make my life better.

I've had just about all that I can take.
I am so tired of making the same mistakes.

We have to always strive to be better than what we are.
And stop being held back in life because of our past emotional scars.

If we turn our lives over to the Lord.
We wouldn't have to worry about others breaking our hearts.

And once we start feeling good about ourselves.
We can experience the goodness of God's wealth.

Each day I give thanks to my Father above.
For His mercy, His grace, and His undying love.

Another's Pain

I remember waking up in the morning, sitting in my car outside a club.
Robbed or everything that I had while I was doing drugs.

Only by the grace or God did I still have my car.
My addiction kept me in a spiritual war.

But this was the life that I chose to live.
I was too busy taking; I didn't know how to give.

I remember the poverty and death I witnessed overseas.
How their situations worsened regardless of their pleas.

You could see the rib bones sticking out on kids in all the places.
And the horrific look that they had on their faces.

The problems that they were having, made me feel ashamed.
It made me feel bad to ever complain.

It's amazing when we compare our problems to others, just how
small they seem
The luxuries that we have, to some can only be a dream.

Even though I lost a few valuables, I was thankful for having my life.
Because I know in my heart that Satan wanted me dead that night.

No matter how close we are to God, problems will remain.
But the best way to see our blessings is to take a look at another's pain.

Walking by Faith, not by Sight

What looks good for the eyes may not be good for soul.
By pleasing our flesh, we may be condemning our soul.

It's amazing how we cause problems in our own life.
By not walking by faith, but walking by sight.

You see, what looks good in the physical shape.
Can have a negative effect on the decisions that we make.

By not walking by faith, but walking by sight.
We constantly find ourselves in a spiritual fight.

There's nothing weak about any of God's creations.
Which is why He gives us a way out or any bad situation.

How many times have we said, I'll believe it when I see it.
Not realizing that a lack of faith is actually what we're dealing with.

And the very things that look pleasing to our eyes.
Are nothing more than spiritual lies.

So instead of asking God to show us what is right.
We're constantly making bad decisions purely by sight.

If we listen we may find ourselves learning an important lesson.
And realize that by walking by sight, we're missing our blessings.

It's time for us to stop falling for the devil's bait.
And stop walking by sight and learn how to walk faith.

That's How We Plant Good Seeds

The greatest loss in life is what dies inside of us while we live.
That's why we can't lose our faith; we have to believe that God is real.

We can't let the trust and love that we have die.
The answer to our problems is not to run and cry.

Whatever we're going through, somebody else is in worse shape than us.
Which should be enough reason to always give thanks for God's
    mercy, grace, and His love.

God knows that some people are just stronger than others.
And that's why He put us all here, to carry each other.

It's easy to turn our backs on problems when we're not involved.
But that's the perfect opportunity to talk about the goodness of God.

There are bad people in this world, but there are good people too.
And it's the good people's job to spread the Gospel/good news.

We have to stop judging others just because of how they look.
And realize that there's more to a story than the cover of the book.

The only way for us to understand what someone is going thru.
Is to take the time to walk in that person's shoes.

God will lead us in the right direction.
If we take a good look in the mirror at our own reflection.

It's up to us to be an encouragement to others during their time of
    need.
In this time and place that's how we plant good seeds.

And it's wrong putting someone down using the Bible as a tool.
We may be able to hide the truth from others, but God can never be
fooled,

If we concentrate more on what's inside of a person instead of what
we see on the outside.
Then just maybe we'll start seeing others with our hearts instead of
our eyes.

People are being persecuted simply because of the color of their skin.
With no consideration of what kind of person they are within.

We often think that how we look on the outside will dictate how we
feel within.
Not realizing that vanity and pride are a gateway to sin.

And our self-esteem always gets caught in the way.
Fooling us to believe that we have no reason to pray.

And that's why we have to break loose from the chains that are hold-
ing us down in our life.
We have to get our lives back on the right track by first accepting
Christ

I don't care if we're at work, school, at a store, or at home.
We have to let someone know that Jesus still loves them, even when
they're wrong.

No matter what it was that they've done before, Jesus will forgive
their sins.
And let them know that it's not where they were that counts; it's all
about where they end.

Jesus loves us so much that He sacrificed His life so that we could live.
He wanted the whole world to know that we were His.

And we should always be willing to help someone when they're in need.
In this time and place that we're in, that's how we plant good seeds.

God's Will

Even though we may feel that we're going through troubling times.
There is one thing that we have to keep in mind.

That God is always keeping a watchful eye.
So we never have to ask the question why.

Why am I going through so much, am I being punished?
In order to fulfill our purpose in life, there are
certain things that we must accomplish.

First we must recognize when we're doing
wrong, and find a way to make it right.
In order to do this, we must follow God's
instructions day and night.

And when we pray, we shouldn't ask for selfish needs.
We should pray that others have the right mind to believe.

And by us praying for blessings for others.
We'll demonstrate the love that we have for our sisters and brothers.

God opposes the ones who act too proud.
For a proud person is too prideful to worship out loud.

And He seems to think His accomplishments are His own.
Those are the ones whom God will disown.

We are nothing without Christ in our life.
Without Him, nothing we do will ever be right.

Everything that we go through, there's a lesson to be learned.
That it's not God's will that anyone burns.

Just Don't Give Up

Laid down full of depression, rotting away in a cell.
Feeling as if I was sitting in the footsteps of hell.

Ready to give in to the devil's request by giving my soul away.
Blaming and cursing God for my life going astray.

Not realizing that I was incapable of understanding God's plan.
Thinking that all my prayers should be
answered simply on my demand.

As I sat there all alone, feeling so self-absorbed.
God asked me, "Are you tired, have you had
enough, or do you think you need more?"

It was my belief that the only way to receive love was to give love.
Being deceived by Satan, not knowing that
God's love is unconditional.

I couldn't believe that God could love a sinner like me.
I didn't think it would do me any good to get down on my knees.

The enemy led me to believe that I was
okay; I didn't need anyone's help.
I could handle whatever problem that I faced all by myself.

But there was something that I just didn't
know because I couldn't understand.
No matter how bad I was, Jesus still loved me, and my life had a plan.

A voice once told me, "Don't give up. One day
you're going to help fill my pews."
"Your testimony of your deliverance is what you're going to use."

So what you think is the end of the world is only the beginning.
Of a lifelong union with Jesus Christ to
encourage others from sinning.

Even when you don't know it, there's a prayer
going out for you at another's request.
Which explains those tight situations that you
came through, one step away from death.

The only way we can have change from bad
to good is to say enough is enough.
And take heed to that voice that's constantly
reminding us to just don't give up.

One

We're just one check from being homeless,
and thrown out on the streets.
One person away from running into the
best friend that we could meet.

Just one word from an argument, that could divide a home in two.
One lie from a heartache, when it was just as easy, to tell the truth.

Just one minute from being fired, and losing everything we owned.
That car, that truck, our family, and even that brand new home.

Just one drink away from a disease which
would cause negative reactions.
All it takes is one snort of cocaine to start
a life plagued by addiction.

It takes one needle in the arm, to be labeled as a junkie.
Throwing away everything that you were
taught or led in life to believe.
If for once we start showing less hate and more love.

Maybe then our blessings will start raining from heaven above.
A man who believes in nothing is a man who has nothing to believe.
Just think of all the things we can accomplish
if we focus on our sobriety.

If we just put all our time into something
positive, other than getting high.
For once in our life we can hold our heads up high.

I have no doubts that blessings will come as
we do things that are constructive.
As long as God is in our lives, we'll always be instructive.

The Light

As I was walking, I felt a presence, but as I
looked around there was no one in sight.
And amidst all the darkness which surrounded
me, I noticed a sparkle of light.

Everywhere I went that light always seemed to follow.
And no matter how dark my life got, that light
always brightened my tomorrows.

I use to think that I was destined to always be alone.
Until that light entered my life and proved that I was wrong.

So now I give great pleasure in just waking up.
But I remember once upon a time in my
life when even that wasn't enough.

I was walking through this life thinking
someone owed me something.
But my selfish reasoning led me to believe that
I didn't have to give back nothing.

Once that light entered into my life, my path became clearer.
And I knew it didn't matter where I was,
that light would always be nearer.

So now when a problem arises, I just get down on my knees.
And no matter what happens, I know He'll never give up on me.

Because I know now that, that light will never leave me in the dark.
And I can honestly say that I love that light
from the bottom of my heart.

So if you ever feel darkness surrounding your life.
Seek Jesus and that darkness will turn to light.

I Have a Plan

I have a plan, to put my life back together.
It starts with me finding the strength within,
to overcome all stormy weather.

I have a plan, to stay clean and sober every day.
And I thank God that I finally realized that it's never late.

I have a plan, to repair all of the bridges that I once burned.
And use the experiences of my past failures
as a growing tool to learn.

I have a plan, to be an inspiration to all of God's lost sheep.
So that maybe I can make a difference, and
end the nightmares in my sleep.

I have a plan, to get a career and have a brighter future.
So that I can stop thinking small, and see the bigger picture.

I have a plan, to dedicate my life to righteous living.
But first I must pray for a heart that has no problem with giving.

I have a plan, to stay free and out of prison.
And start excepting responsibilities for my
actions, regardless of the reasons.

I have a plan, to voice out to the children of the streets.
That through Jesus, not acts of violence are
they able to overcome being weak.

I have a plan, to strive forward and never look back.
To stand firm on all my strong points, instead of those I lack.

I have a plan, to take just one day at a time.
Because I know that recovery doesn't have a
chance unless it's made up in the mind.

I have a plan, to be as humble as I can be.
And be obedient to my God with the plans He's chosen for me.

Because, no matter how I plan, according to God's words.
None of these things can ever happen, if it isn't His will first.

SHHH!

Shhh, can you hear it? It's softer than a whisper.
They're celebrating in heaven; another soul has been delivered.

Shhh, can you hear it? The angels are lifting their voices in praise.
It's time for all believers to pack their bags;
the Lord has chosen this day.

For all tongues to confess, and all knees to bow.
To be held accountable for all the things we've done,
for that is what judgment day is all about.

There is a place for believers; it's called heaven,
Where all the streets are paved in gold.
And if you are a non-believer, this day I feel sad for your soul.

Everything that is happening was written in His word.
He has sent many messengers out to ensure that His truth be heard.

The only way to salvation is through Jesus Christ Himself.
If you have Him in your life, you no longer fear death.

Within My Heart

How can I find forgiveness within my heart!
After it has been ripped, stripped, and torn apart.

When I've given my all, with nothing to show in return.
Left only with loneliness and grief over a feeling that burns.

How can I find forgiveness within my heart!
When it wasn't there from the start.

Continually hurting others with no regret, with no regard.
Fooling myself into thinking that I'm hard.

Deep down I was just as scared as a small child alone in the dark.
Mad at the world because I haven't lived up to my part.

To add fuel to the fire, I put drugs and alcohol in my system.
Which led me to believe that I was the victim.

All along I was running, lying, and hiding from myself.
Destroying everything in my path until there was nothing left.

Too foolish to admit my wrong, too proud to ask for help.
Until I found myself looking at the face in the
mirror, one step away from death.

That's when I saw my life flash right before my eyes.
Which caused me to fall down on my knees and break down and cry.

It was then my heart changed, and I asked
forgiveness from my heavenly Father.
Only through Him was I able to see another tomorrow.

Stop Pretending to Be Something You're Not

Why do we pretend to be something that we're not?
Instead of just being satisfied with what we got.

And why must we put our life out on the edge.
Just because we're worried about what's being said.

This is our life, and we get only one chance to live it.
And that means that there are things that we don't have to deal with.

In a world of give and take, we have to always be
alert to the predators in our midst who form.
An endless search for the perfect victim
to lay down their wrath upon.

With the use of either by force, so while we're flashing,
remember we can draw the wrong attention.
Just because we couldn't accept who we are with contention.

Not everyone can be at the front of the line.
So be thankful for what you have; you too will have your time.

And pray for God to give you a humble spirit.
And be thankful for your life, while you live it.

You owe no one any kind of explanation.
Just remember there are others out there in worse situations.

I'm going to be me, regardless of whom I'm around.
Because I know that my faith will keep
me standing on solid ground.

You Can't Go Wrong with God

You can't go wrong with God in your life.
He's going to mold you and transform you until you get it right.

Only God is perfect, but we can give our best.
As long as we don't give up and settle for less.

You can't go wrong with God in your heart.
In fact, you're going to be determined to do your part.

Remember how twisted your life was before.
You took heed to God's voice and opened up your doors.

You can't go wrong with God in your mind.
He'll change the way you think, and you'll
leave your troubles behind.

Anything that we do, we have to involve God in our plans.
And realize we're in no position to make any demands.

You can't go wrong with God in your sights.
Even in your darkest moments, He'll give you a sparkle of light.

We'll give up on ourselves, yet His love will never die.
And every time I think back to what He delivered
me from, I just break down and cry.

The Pulpit Is No Place for a Star

All saints of the world need to be on one accord.
By being strong and vocal while standing up for the Lord.

We are God's warriors, we should be
turning this world upside down.
Converting sinners to saints wherever they are found.

God's people shouldn't know the meaning of the word quit.
And they should always be mindful of whom
they are hanging around with.

We have to first learn to obey God, and do His bidding.
By letting Him be a part of our everyday living.

Christians should never be afraid of walking outside their homes.
With God by our side, we're never alone.

We were put here to be fighters for the Lord.
So why are we depending on others to do our part.

When we were serving the devil, we gave Him our best effort.
So how much more is God's love worth.

It's time that we all get together and start saving souls.
Without unification a church can never grow.

Most of our leaders have their hands in the cookie jar.
Using the pulpit as a stage to become a star.

Twisting the words of the Lord for their own belief.
Oh, what a sad day it'll be for them when they face the heat.

If the message being spoken is not about Christ.
Then the sermon itself just ain't right.

So whenever you see a saint traveling down a weak road.
Pray that our Lord Jesus strengthen their souls.

There's no reason that a saint should ever be alone.
And there's enough power in God's words to keep us all strong.

Now is the time saints that we all should stand on one accord.
No matter what we're up against, we're protected by the Lord.

So remember saints, that this is no time to be a star.
As long as good and evil is caught in war.

God wants humble saints to spread His words.
That's the only way His true message can be heard.

Just Because

Just because we go to church that doesn't make us no saints.
Especially if we don't know the difference between
what we can do and what we can't.

Remember, Satan is going to always try
to put someone in our midst.
So be careful what you ask for because you may regret what you get.

So if you see someone showing compassion,
love, and encouraging others.
It's safe to say that you're dealing with true
Christian sisters and brothers.

But always be mindful when you see a crooked smile.
Trying to convince you that living a Christian life is out of style.

Satan knows the Bible very well inside and out.
That's why He's constantly trying to fill our minds with doubt.

And He doesn't care whom He uses on the way.
He's willing to do whatever it takes to set us astray.

Because He knows that in unity we have a solid foundation.
After all we are God's number one creation.

Just because we go to church, that doesn't mean we're living right.
Worshipping God in the daytime and then partying all night.

It's impossible to serve two masters at one time.
That's why it's important that we make up our minds.

And if you're having problems understanding how we must live.
Read the Bible for your answers; its whole purpose is to give.

So put your trust in God, and He'll be there to see you through.
There's no limit to the blessings that He has in store for you.

A True Friend

When I am in need of a true friend. Someone I could always confide in. Someone who'll never raise their hand. Other than to teach me to do the best I can.

There are times when my mind is filled with thoughts. Of the peace of mind that you have brought. Knowing you would never be bought. And your love for me will never come to a halt.

I'll always hold my head up high. Because there are certain things I cannot deny. And I will never have to ask the question why. As long as I have you by my side.

When I am tired, and I am cold. You're always there to lighten my load. You fill me up body and soul. With you by my side, my life is whole.

When I am weak, and I am scared. That is the moment that I seek your word. You let me know that my prayers are heard. Because a peace of mind always seems to come first.

When I am hungry, and I am broke. You're always there to give me hope. No matter what happens I now know. As long as I have you in my life, I will always cope.

With you on my side, nothing would be too hard. And you and I would never be apart. You'll always be special in my heart. So that is why I must say, I truly do love you, Lord.

Lost in Eternity

~~~~~~~~

Small L.I.E.

We may be on the verge of a blessing, and
we do something to turn it away.
It could even be something as small as a lie that
may have come out of our mouths that day.

I don't believe that there is such a thing as a small lie.
And if you do, can you please explain to me why.

What starts off as small, will soon begin to get bigger.
It could even change the course of your life
when the truth would have been clearer.

One sin is not smaller nor greater than another.
And it is foolish to think that you can lie one
day, and it'll be forgotten by tomorrow.

It's a good feeling when you are honest and always tell the truth.
Of course hurting someone emotionally is
something we never want to do.

But it's best to tell the truth, and then deal
with the consequences afterward.
Even though there are times that the truth may seem too hard.

You must know that the only way to get to
heaven is through Jesus Christ.
And that's not going to happen if there's dishonesty in your life.

God hates a liar, just as much as He hates a killer.
So how can you not be truthful, and call yourself a believer.

When God's spirit is in you, your vision
becomes clearer, and you are able to see.
That if you think it is right to tell a small lie,
then your soul will be lost in eternity.

Giving Thanks

As I was walking down the street one day.
I noticed that the pigeons were swooning in a peculiar way.

The trees were swaying from side to side.
And the ants stood at attention right before my eyes.

The birds were chirping all in unison.
As the flowers were blooming beneath the sun.

I saw a pack of dogs, and they were all sitting on hind legs.
And even though I had food in my hand not one of them begged.

Then I noticed that three cats were meowing.
It was as if they were trying to sing.

I also looked in a pond and saw ducks swimming in a row.
It was as if all these creatures were putting on a show.

I couldn't understand what was happening right before my eyes.
How the creatures were behaving from the land to the skies.

Then God said, "Why are you confused?
The answer shouldn't be too hard"
"You should know that every living thing worships the Lord."

So what makes us think that we know everything.
When we don't even know what tomorrow will bring.

So do like the ducks, ants, birds, dogs,
pigeons, cats, flowers, and the trees.
And stop what you are doing and get down on your knees.

Give thanks, if for nothing else, for the very air that we breathe.
Because He shows us every day just how
much He loves you and me.

Let's All Join Hands

Let's all join hands and express our deepest love.
Let's all join hands and share in prayer to our Father up above.

There are times when we all give up and feel as though we can cope
no longer.
But the true doctrine of the Lord will make us become even stronger.

For it is written in His word that He truly does love us, and He'll
never leave us nor forsake us.
Thru sickness, hunger, grief, and loneliness, we will conquer all, as
long as we abide in His word from up above.

Let's not lose our memories and forget the great blessings He has
brought our way.
Let's all reflect on trouble times when a solution seems impossible,
then Christ came to our needs and produce a better day.

So let's all join hands and not forsake the Lord for the pleasures of a
sinful world that one day will cease to exist.
Whether we live by the way of the Lord or by the sins of the world,
just remember one day before the Lord these ways we will all
have to admit.

The time is now that we all stop guessing, and know the true light
of salvation.
So let's join hands and understand the Lord's word, and pass it on
from generations to generations.

If This Life Was Only a Dream

If this life was only a dream, and life after was reality, and no one
knew nothing of it.
And then someone was to pass away, and his dreams were completely
over with.

I know that when we dream our eyes are closed, but there are things
that we still may see.
So who knows maybe that last breath that we take, may open up a
gateway toward true reality.

Whom are we to say whether or not someone lived a life of fantasy,
or of nightmare.
For our dreams are so complex that even the wisest of men are left
unaware.

And if this life was truly a dream, I would have but one prayer to ask.
And that would be to awaken me from this nightmare, even if it does
mean that I must pass.

We must not question what happens in our lives, for control is not
in our hands.
For if we had power overall, men would have created men.

From Darkness to Light

Out of the darkness and into the light.
Sprouted a new beginning into our life.

The things of the past become the things of old.
Washed away in the sea of forgiveness to be forever untold.

With belief and faith, patience comes forth.
And the mistakes of yesterday are nevermore.

How far are we willing to go to achieve our goals.
Of reserving our spot in heaven by saving our souls.

Out of the darkness and into the light.
Sprouted a new beginning into our life.

The key is to never give up, a change will come.
Just remember that what's important to
you may mean nothing to some.

Out of the darkness creepy things will crawl.
Which is their time to devour us all.

But through the light, we're able to fight off their attacks.
And leave them standing right in their tracks.

Since the beginning, darkness has been at war with the light.
Deploying it's evil within the cover of the night.

But no matter how dark it gets, whenever the light comes.
Darkness has no other choice but to pack up and run.

There's No Place Like Home

Often a baby bird leaves immaturely from his nest.
And leaps to his doom after giving his best.

Even though his little wings were flapping
one hundred miles an hour.
He was still too young and without enough power.

For some reason we always think that the grass
is greener on the other side of the road.
And the streets are like heaven painted and embedded in gold.

It's these delusions of grandeur which cause us to fall.
By stripping us of all our dignity and making us feel small.

For each mistake we make, there is a lesson to be learned.
All we have to do is have patience and wait for our turn.

But every time we get blessed, we want to give credit to ourselves.
Until everything is gone, and we're all alone
with nothing to show that's left.

And those so-called friends didn't offer to lend you a hand.
Instead when they were around, they made you feel less than a man.

You were hungry and homeless, but what could you do?
Even the pigs that you tended are better than you.

It's time for you to lose that foolish pride and those ignorant ways.
And call on your heavenly Father before it's too late.

Stop asking yourself, what happened? Where did I go wrong?
Because Jesus never left your side; He just wanted you to see for
yourself that there's no place like home.

He's Worthy to Be Praised

Growing up, going to church, I had no understanding.
That at that particular time in my life I was
experiencing a spiritual beginning.

I did not know how blessed I was to be a part of God's family.
Blinded by what the world had to offer
and what I just couldn't believe.

That without God nothing would exist,
so He's worthy to be praised.
We need to let Him know how much we
appreciate Him each in every day.

Not only does He fulfill my needs, He puts comfort in my life,
and His hand is always there to protect me,
even when I'm not living right.

No matter how my day ends, He always seems to wake me up.
All He wants is for me to love Him and that's not asking too much.

Because He knows that if you love Him
you're going to follow His word.
And bring others close to Him by ensuring
that His message be heard.

Love is the key to holding together a solid and strong foundation.
And if your love is true, then your actions require no explanation.

There are many members in church that are on their way to hell.
If for no other reason, for those lies they're trying to sell.

I don't care how good you look on the outside;
God grades you from within.
And your actions alone will tell the story of
whether or not you're living in sin.

God is a respecter of no man; He can use anyone to teach His way.
And He'll never stop loving us. That's
why He's worthy to be praised.

Index

About the Author

I was raised in the southern part of Dallas. As a child I was always sickly due to breathing problems. I eventually started playing football which I was very good at. I received a scholarship to play football at numerous schools. I served in the U.S. Navy and was honorably discharged. I experience some bad moments in my life at an early age. And if it was not for God's loving, unforgiving love, I would not be here today. I thank God and my wife Iris of twelve years who have always been an inspiration in my life.

www.ingramcontent.com/pod-product-compliance
Lightning Source LLC
Chambersburg PA
CBHW031323290526
45784CB00014B/878